How to Install Simple Water-Saving Irrigation Systems in Your Yard

GREYWATER GREEN LANDSCAPE

Laura Allen

Storey Publishing

The mission of Storey Publishing is to serve our customers by publishing practical information that encourages personal independence in harmony with the environment.

Edited by Deborah Burns
Art direction and book design by Jessica Armstrong
Text production by Erin Dawson
Indexed by Samantha Miller

Cover photography by © Michael Dorman (front) and © Rebecca Louisell (author)
Back cover illustration by © Steve Sanford
Interior photography by © Michael Dorman: vi, viii, 4, 16, 32, 38, 39, 44–46, 47 top, 54, 56, 65, 66, 74, 78, 84, 90, 92 left, 93, 95, 96, 99–101, 103–105, 107–111, 113, 115, 116, 126, 128, 129, 132, 133 top, 134, 135 left, 140, 143–145, 147, 160, 182
Additional photography by © Advanced Waste Water Systems, 157; © Alan Hackler, 92 right; © anakeseenadee/iStockphoto.com, 159; © Arterra Landscape Architects, 173; © Art Ludwig, from *Create an Oasis with Greywater*, oasisdesign.net, 6; © blueskyline/iStockphoto.com, 28; © Claus Alwin Vogel/iStockphoto.com, 12; © Evergreen Lodge at Yosemite/Kim Carroll Photography, 10; © GWA Bathrooms and Kitchens, NSW, Australia, 60; © iammai/iStockphoto.com, 48; © jamroen/iStockphoto.com, 31; © Jeremiah Kidd, 176, 177; © Josh Lowe, 141; © Laura Allen, 11, 19, 72, 97, 136 right, 138; © Laura Maher, 87 bottom; © Leigh Jerrard, Greywater Corps., 13, 29, 58, 79, 86 top, 87 top & middle, 133 bottom, 136 left, 153; Mars Vilaubi, 135 right; © Regina Hirsch, 70; © Richard A McGuirk/Shutterstock, 9; © Rocky89/iStockphoto.com, 47 middle; Courtesy of San Francisco Public Utilities Commission, 76; © Sergio Scabuzzo: vii, 117 bottom; © Ty Teissere, 64; © Yarygin/iStockphoto.com, 80
Illustrations by © James Provost, 15, 17, 19, 20–25, 40, 41, 50, 67 bottom, 86 bottom, 88, 89, 93–95, 97, 98, 106, 108, 110, 117, 125, 126, 130, 131, 134, 137, 146, 149, 155, 163, 165, 175, 178–180, 182, 183 and © Steve Sanford, 2–3, 42, 47, 67 top, 81, 83, 85, 102, 114–116, 119, 121, 129, 139, 161, 168, 171

Portions of this book were excerpted from *The Water-Wise Home* by Laura Allen (Storey, 2015).

Storey books are available for special premium and promotional uses and for customized editions. For further information, please call 800-793-9396.

Storey Publishing
210 MASS MoCA Way
North Adams, MA 01247
storey.com

Printed in China by Toppan Leefung Printing Ltd.
10 9 8 7 6 5 4 3 2 1

Library of Congress Cataloging-in-Publication Data

Names: Allen, Laura, 1976–
Title: Greywater, green landscape : how to install simple water-saving irrigation systems in your yard / Laura Allen.
Other titles: Grey water, green landscape
Description: North Adams, MA : Storey Publishing, 2017. | Includes bibliographical references and index.
Identifiers: LCCN 2016051368 (print) | LCCN 2016052283 (ebook) | ISBN 9781612128399 (pbk. : alk. paper) | ISBN 9781612128405 (Ebook)
Subjects: LCSH: Graywater (Domestic wastewater) | Water reuse. | Lawns—Irrigation. | Residential water consumption. | Water consumption.
Classification: LCC TD429.A4234 2017 (print) | LCC TD429 (ebook) | DDC 627/.52—dc23
LC record available at https://lccn.loc.gov/2016051368

To the greywater installers, educators, and users,
and especially to Greywater Action

Contents

PART 2

Building Your Home Greywater System 79

Preface

In August 2012 I sat in a radio studio talking about greywater with my former plumbing teacher — and, at the time, Oakland's senior inspector — Jeff Hutcher, on American Public Media's *The Story.* When I met Jeff in 1999, I never would have imagined this moment. I was a student in his residential plumbing class, seeking to learn hands-on skills to build sustainable water systems. At the time, he was horrified to hear about my "Frankenstein" greywater setups and refused to answer questions about my then-illegal plumbing systems in class (he was, after all, a city inspector). Now he and I work together to streamline permits and facilitate legal reuse of greywater.

For the past 17 years, I've designed and built simple residential water reuse systems: greywater systems, rainwater catchment, and composting toilets. Once my friend and housemate Cleo Woelfle-Erskine and I cut into our home's plumbing to channel the shower water outside. I couldn't imagine ever again letting this good irrigation water escape to the sewer. We taught our friends and wrote about the how and the why of it. Our group, the "Greywater Guerrillas," grew

Greywater Action Los Angeles in 2016. Left to right: Ty Teissere, Laura Allen, Cris Sarabia, Laura Maher, Sergio Scabuzzo

out of these projects. Later, we worked on an anthology, *Dam Nation: Dispatches from the Water Underground,* which placed greywater reuse, rainwater catchment, and composting toilets in the larger political context of water issues around the globe.

Some genuine plumbers joined our group: Christina Bertea, the first woman admitted into Local 159, Plumbers and Steamfitters Union, and Andrea Lara, then an apprentice. With their involvement, we honed our skills and revamped our designs. Andrea, Christina, and I taught dozens of hands-on workshops all over the Bay Area and southern California. As our state entered a multi-year drought, we couldn't keep up with the demand. Every workshop filled up, along with the wait list. I gave talks at green-living festivals, universities, churches, and even high schools.

Since all our work was illegal, according to state plumbing code at the time, we became involved in changing the code. In 2009 the State of California overhauled its greywater code, making many greywater systems legal. That same year we renamed our group "Greywater Action: For a Sustainable Water Culture," to represent our goals and strategies to a diverse audience.

We continue to teach hands-on workshops, as well as trainings for professionals who want to offer these services to their clients. In our one-week class we teach people theory and hands-on skills, culminating with the participants installing a real system on the last day of class. Over a hundred people from across the U.S. and Canada have graduated and now champion greywater in their communities. Their systems, businesses, and workshops are the ripple effects of their training.

Our work expanded from the Bay Area to Southern California, where I lived for the past few years, with new members, bilingual (English/Spanish) trainers and materials, and partnerships with forward-thinking water agencies.

Now I live in Oregon and am adapting greywater into a new climate region.

I hope this book helps you tap into your own greywater resources and grow a beautiful, productive landscape.

Brian Munson on left, Christina Bertea on right

PART 1

Planning Your Home Greywater System

If you're like most people, you wash clothes, take showers, and run water down the sink. Why let this good irrigation water go to waste? Surely you've heard the expression, "Don't throw the baby out with the bathwater." Now, let's stop throwing out the bathwater!

This part of the book — chapters 1 through 7 — takes you through the initial design and planning steps for building a greywater system. A successful system is tuned to match your water usage, your home's plumbing, and your landscape. Here, you'll learn how to estimate how much greywater your home produces, which greywater sources (called fixtures) you can tap into, and how to test your soil and size the mulch basins (to receive the greywater in your landscape). You will also learn about the types of plants that grow well with greywater and which soaps and products are safe to use. In addition, we'll look at some basic health and safety considerations and regulations for greywater. After you've done the preparation work, Part 2 — chapters 8 through 11 — will take you step-by-step through the installation of various systems.

Don't forget to start with efficiency. Before planning your greywater system, make sure your home is water-efficient. Fixing leaks and upgrading fixtures can reduce indoor water consumption by around 35 percent. See the Resources section for more information.

CHAPTER 1

Greywater Systems 101

The first step to design and install a greywater system is to understand your options.

These include the different types of systems, their general costs, their strengths and limitations, their benefits, and their health and safety considerations. This chapter provides the foundation you'll need for the details and calculations in the upcoming chapters.

IN THIS CHAPTER:

What Is Greywater?

Greywater is gently used water from sinks, showers, baths, and washing machines; it is not wastewater from toilets or laundry loads containing poopy diapers. Plants don't need clean drinking water like we do! Using greywater for irrigation conserves water and reduces the energy, chemicals, and costs involved in treating water to potable quality.

Reusing water that we already have is a simple and commonsense idea. Just use "plant friendly" soaps (those low in salts, and free of boron and bleach), and you have a good source of irrigation water that's already paid for.

Greywater systems save water and more. They can extend the life of a septic system, save time spent on watering, act as "drought insurance" (a source of irrigation during times of extreme water scarcity), and encourage the use of more environmentally friendly products. They also use less energy and fewer chemicals than other forms of wastewater treatment.

Home using greywater for irrigation with simple laundry-to-landscape and gravity-fed systems

A WATER-EFFICIENT HOME AND LANDSCAPE

Before you start planning and constructing your greywater system, be sure to make your home and landscape as water-efficient as possible. Leaks waste an average of 14 percent of total home water use. Toilet and irrigation system leaks are often hidden and go undetected. A simple "mini makeover," such as switching out water-guzzling fixtures and appliances for efficient models, can lower total household water use by up to 35 percent.

Equally important is making your landscape water-smart. Plants that aren't able to be irrigated with greywater (or rainwater) should be those that are adapted to your local climate and are able to thrive without potable irrigation.

This book guides you through the design and installation of several types of greywater systems, but to maximize the full range of your water resources you'll also want to incorporate rainwater harvesting into your overall landscape design. See Resources for more information.

Water Savings from Greywater

You can expect to save between 10 and 20 gallons per person per day (or more) from a greywater system, though this number can fluctuate greatly. Studies estimate savings of between 16 and 40 percent of total household use. How much you actually save depends upon how much you currently irrigate, whether you use greywater on existing plants or you plant new ones, and how many greywater sources you can access. One study in Central California found an average household savings of 15,000 gallons per year after the greywater system was installed (see Resources for more information). For tips on how to maximize water savings with your greywater system, see Maximize Water Savings on page 46.

Cost of Greywater Systems

Materials for simple greywater systems typically cost a few hundred dollars. If you're handy, you can install a system yourself in a day or two. Professional installations range from $700 to many thousands of dollars, depending on the type of system and your site. We'll discuss more details about specific system costs in chapter 6.

Types of Greywater Systems

There are many types of greywater systems, ranging from simply collecting water in buckets to fully automated irrigation systems. I'll group them according to their relative level of complexity and briefly explain how they work.

PROS OF GREYWATER SYSTEMS

- Greywater is produced every day, all year long, and is a reliable source of irrigation.
- Simple systems recycle tens of thousands of gallons a year for a relatively low cost.
- Systems take up little space; often, all the pipes are buried and invisible.
- It's easy to irrigate fruit trees, shrubs, and large annuals and perennials.
- It's an automatic system, saving time and ensuring plants get watered.
- It reduces wastewater going to the sewer or septic system.

CONS OF GREYWATER SYSTEMS

- Accessing greywater may be challenging, depending on how your house and landscape are designed.
- Greywater reuse is not yet legal in some states.
- Requires use of "plant-friendly" products in the house.
- Small plants, or plants spread out over a large area, are more difficult to irrigate with the simplest systems, though pumped and filtered systems will work.

"LOW-TECH" SYSTEMS for irrigation are the lowest in cost, simplest to install, and easiest to obtain permits for. Common types include **laundry-to-landscape (L2L)** and **branched drain** systems.

"MEDIUM-TECH" SYSTEMS for irrigation incorporate a tank and pump to send greywater uphill or to pressurize it for drip irrigation.

"HIGH-TECH" SYSTEMS are used for automated drip irrigation or toilet flushing in high-end residences and larger-scale commercial or multifamily buildings.

Hooray for the Washing Machine!

Washing machine water is typically the easiest source to reuse; you can direct greywater from the drain hose of the machine without cutting into the house's plumbing. A washing machine has an internal pump that automatically pumps out the water and can be used to direct greywater to the plants.

No-Fuss Gravity Systems from Showers and Baths

Showers and baths are excellent sources of greywater, though accessing the drainpipes may be challenging, depending on their location. A diverter valve placed in the drain line of the shower allows greywater to be diverted to the landscape. Gravity distribution systems are usually cheaper and require less maintenance than pumped systems, and distribute greywater through rigid drainage pipe. Greywater flow is divided into multiple irrigation lines to irrigate trees, bushes, vines, or larger perennials via mulch basins (see page 42).

Greywater Pioneer

Art Ludwig

Art Ludwig, affectionately called "The Greywater Guru" by many, is an ecological systems designer with 35 years of experience in the field.

He has studied and worked in 22 countries, authored the books *Create an Oasis with Greywater* (an excellent resource containing his decades of greywater experience) and *Builder's Greywater Guide*, and produced the "Laundry to Landscape" instructional video. He created the first plant and soil "bio-compatible" laundry detergent, designed to break down into plant nutrients (see Resources). Art's most popular greywater inventions include the laundry-to-landscape and branched drain greywater systems. His policy work has greatly improved regulations in California, New Mexico, Arizona, and elsewhere.

Greywater-irrigated landscape

What advice do you have for people wanting to install a greywater system?
Choose the simplest possible approach and implement it as well as possible.

What's your all-time favorite greywater system?
My favorite greywater system is my first branched drain system. I thought it would fail in a day, but it didn't. If fact, I couldn't get it to fail no matter what I did. Even after removing the screen in my kitchen sink and pushing compost scraps down the drain, the sink trap clogged daily, but everything that made it past the trap flowed out to the mulch basin via a free flow outlet, with no odors, no clogging, and no problems.

Pumped Systems: Filtered and Unfiltered

Pumped systems push greywater uphill or across long distances. Greywater is diverted into a **surge tank**, from which it's pumped to the landscape. Adding a filter allows greywater to be distributed through smaller tubing, increasing the potential irrigation area but also increasing the cost and maintenance of the system.

Health and Safety Considerations

A properly designed greywater system is safe for you and your family as well as the environment. As you design your system, keep the following considerations in mind:

- Greywater is not safe to drink or ingest.
- Greywater can harm aquatic ecosystems.
- Avoid direct contact with greywater; your system should not create a puddle, pond, or any standing water, which creates a hazard to unsuspecting children and a place for mosquitoes to breed.
- When watering food plants, don't let greywater contact edible portions of the food (to avoid direct ingestion). Don't irrigate root vegetables. Use subsurface irrigation for growing edibles near the ground.
- Never use greywater in sprinklers where the spray could be breathed in.

Polluting Aquatic Ecosystems

The nutrients found in greywater are good for your garden but harmful to lakes, streams, and oceans. In the garden, the nutrients are fertilizer, sent to your plants each time you do laundry. If, instead of soaking into the soil, greywater runs into the storm drain and out to a river or bay, the nutrients pollute the water, feeding algae and robbing oxygen from aquatic organisms. This is similar to how the gigantic "dead zones" are created in our bays and oceans, usually by fertilizer runoff and sewer overflows. Don't let your greywater contribute!

GREYWATER OR GRAYWATER?

People often wonder why greywater is spelled two different ways. All around the world greywater is spelled with an "e," except by a few groups in the U.S., mostly regulators who write state codes (though some states, like Washington, use the "e" spelling). I like the "e" spelling to intentionally connect greywater to the global movement around water.

IS GREYWATER A HAZARD TO YOUR HEALTH?

Some people hold the mistaken belief that greywater is hazardous, on par with sewage. There are many studies on the quality of greywater with respect to public health. Here are some key points:

- → Greywater is not safe to ingest. The quality is lower than drinking water quality. **So don't drink greywater!**
- → Greywater-irrigated soil is not safe to ingest, nor is soil irrigated with tap water. The City of Los Angeles conducted a study comparing greywater-irrigated soil and non-greywater-irrigated soil. Both were found to be unhealthy to ingest.
- → Many studies have found fecal indicator bacteria present, demonstrating the potential for greywater to contain fecally transmitted pathogens. A problem with using indicator bacteria, like fecal coliforms, is that the bacteria are present in all mammals' feces and can multiply on their own, resulting in inaccurate estimates of fecal matter in greywater. Other studies tested greywater for a cholesterol found only in the human gut, unable to grow on its own, and found the levels of feces reported in the other studies using indicator bacteria were 100 to 1,000 times too high. Does this matter? Some people still believe greywater isn't much cleaner than sewage and don't know that the methods used to test greywater were grossly inaccurate.
- → Few studies have found actual pathogens in greywater. Testing for specific pathogens is expensive, and pathogens are found only if someone in the home has the illness and germs get into the water. Some studies that tested for pathogens didn't find any, while others found common pathogens like *Cryptosporidium* spp. and *Giardia* spp., which are also found in most surface waters in the U.S. When people are sick they infect others by living together, directly touching, and sharing dishes; the risk of infection through a greywater system is extremely low.
- → There are no documented cases of illness resulting from greywater, while there are over three million documented cases of illness each year (just in the U.S.) from recreational contact with water contaminated by legal sewage treatment plants that overflow.

Is greywater hazardous to your health? No. Most likely greywater use improves public health by reducing waterborne illness since the only people potentially exposed already live together and greywater is kept out of sewer systems that often fail.

Hazardous chemicals should never be used with a greywater system (or anywhere); they will contaminate your yard. When sent to the sewer plant they usually aren't removed either, and will end up in a river, lake, or ocean.

Greywater in Freezing Climates

Live somewhere chilly? Maintaining a greywater system in freezing conditions requires additional planning and precautions:

- Gravity systems should drain completely. Standing water in the pipes could freeze and create a blockage, or potentially burst the pipe. Meticulously maintain proper slope throughout the entire system to ensure complete drainage.
- Do not allow any standing water in lines from pumped systems. Ensure greywater will drain out or drain back into the tank.
- In a pumped or L2L system: If it's logistically difficult to prevent standing water in the line, create an automatic bypass at the beginning of the system. If the main line freezes, water will be forced out the bypass; for example, a tee fitting with a tube running high enough up so greywater doesn't exit unless the line is blocked (if the tube is too short, greywater will come out like a fountain).
- Greenhouses irrigated by greywater (see page 167) can produce food and greenery all year long.

With proper design and installation, greywater can be used successfully in freezing climates.

- Shut off the system (and drain down any places with standing water) until irrigation is needed. Install a drain-down valve at the low point of the system to empty the pipes for winter. Use a tee with a ball valve at the lowest point. Close the valve when using the system and open it to drain the line. **Note:** Shutting off the system may be unnecessary, even with freezing, snowy weather. The warmth in greywater can keep lines open and the ground biologically active (see Cold-Climate Greywater: Evergreen Lodge on page 10).
- Consider a toilet-flushing system if there is no irrigation need. (See Using Greywater Indoors: Toilet Flushing on page 60.)

Cold-Climate Greywater

Evergreen Lodge

Evergreen Lodge (see Resources), in the mountains of Yosemite, California, recycles nearly two million gallons of water annually, thanks to the work of Regina Hirsch of Sierra Watershed Progressive and the lodge's environmentally conscious leadership.

Over the past few years, Regina and her crew installed dozens of systems at the lodge, both simple gravity-fed branched drain systems (in 55 guest cabins) and large automated systems from the commercial laundry and staff dorms.

"The system is a big win for the lodge," reflects owner Brian Anderluh. "We were able to take greywater out of our septic system and use it instead for our landscape beautification project, without requiring any more fresh water from our well."

Regina, an ecologist turned landscaper, notes, "Plant productivity at Evergreen Lodge has increased three-fold since we installed these systems, and soil biota is on the rise."

Even though the lodge gets an average of 30 inches of snow annually, the systems operate problem-free all winter long. The team carefully designed each system so that no standing water remains in the lines to freeze, and the soil remains biologically active from the warmth of greywater.

"Branched drain systems far exceed the pumped systems; however, controller-based valved systems have been reliable for nearly six years with high demands due to drought conditions," notes Regina.

One system repurposes an existing irrigation system. They removed the emitters and sent filtered greywater through the ¼-inch tubing to landscaping in the main courtyard.

Unexpected Benefits

Regina has been monitoring the systems since 2009. "We've seen some really exciting results. The soil is being decompacted by the mulch basin systems. When we first began testing, the soil was so hard we could barely insert the compaction meter rod in an inch; now we are reaching over 62 inches down. Even though many different people use these cabin shower systems and put their different products into the greywater, we have not seen any problems with salt buildup in the soil. And the biggest surprise benefit is the greywater-irrigated plants appear to be fire-resistant!"

In 2013, California's third-largest wildfire (the Rim Fire) burned over 250,000 acres in the Sierra Nevada mountains, and burned within feet of Evergreen Lodge. Regina reflects, "The area was evacuated from August until October, with no irrigation or greywater production. When we finally got in I expected all the plants to have died. Out of 1,500 plants, we lost 90% of the plants irrigated with potable water, but less than 5% of greywater-irrigated plants had died." This is especially significant since two-thirds of all the irrigated plants at the lodge were on greywater.

"The systems at Evergreen Lodge have been so successful, the owners of a new lodge (Rush Creek) being built nearby decided to use 95% of the greywater generated for irrigation. In fact, this new lodge will not use any potable water for irrigation."

Regina adds, "Placing these systems in public view, especially when people are on vacation, is a powerful educational tool. Kids love going out and seeing the water from their shower draining to the trees outside, and the shower rinse water at the outdoor pool flows directly to mulched swales, growing a garden before their eyes. It's so easy to understand what works when it is simple, effective, and in balance."

Greywater from a cabin flows by gravity to irrigate nearby plants.

CHAPTER 2

Greywater Sources and Plumbing

You don't need to be a plumbing expert to install simple greywater systems, but you will need a basic understanding of the drainage plumbing system of your home.

Most importantly, you'll need to determine a suitable location for your greywater diverter valve — before the drain pipe has combined with the toilet. This chapter teaches you these plumbing basics so you can successfully divert the greywater, either yourself or with help from a plumber. I'll also discuss some specific considerations with kitchen sink water, as well as undesirable sources of greywater. We'll end with a summary of greywater sources, noting the pros and cons of each, so you can identify which are a good match for your site.

IN THIS CHAPTER:

- → Identify Your Greywater Sources
- → Your Home's Drain, Waste, and Vent System
- → Kitchen Sink Considerations
- → Greywater-Ready Building Construction
- → When to Call a Plumber
- → How to Locate the Diverter Valve
- → Undesirable Greywater Sources
- → Summary of Greywater Sources with Pros and Cons of Each

Identify Your Greywater Sources

If you're hoping to use greywater from your existing home, you'll find that some greywater sources are easier to access than others. If you're building a new home, you have much more flexibility as to which sources you tap into — and where. Diverting greywater from drainpipes often requires installing a diverter valve (called a 3-way valve) that enables you to switch the water flow between the drain line (leading to the sewer or septic system) and the greywater system. Diverter valves can be operated manually or remotely, via an electrical switch. First we'll look at the primary potential sources for greywater in a home, then we'll discuss the details of tapping into a home's drain system.

Clothes Washers

Washing machines offer the easiest source of greywater to reuse. The machine's internal pump pushes greywater out through the machine's drain hose; from there you can reroute it to the landscape without changing the existing drainpipes. I've worked on hundreds of laundry greywater systems, and they're consistently the easiest and simplest of the greywater options.

In most homes, a greywater pipe begins its route toward the landscape by exiting the laundry room through the wall or floor. Think about how you could send a new pipe from your washing machine out to the landscape. Can it go out through the wall or down into a crawl space and then outside? If your house has a concrete slab foundation and your machine is in a room without exterior walls, the only way to send the water outside is to run the pipe through another room in the house, perhaps hidden under shelving or along a baseboard.

It is easy to live with a laundry greywater system. There are several commonly available greywater-compatible detergents that allow you to safely irrigate plants with the

Laundry-to-landscape system from a stacked washer/drier unit. Valve on the right directs greywater flow to either the sewer or the landscape.

greywater from regular laundry loads. For times when you want to use bleach or wash soiled diapers, just turn the valve located next to the washing machine and redirect the water to the original drain.

Showers and Baths

Showers and baths produce large volumes of good irrigation water, although diverting it to the yard can be tricky, particularly in existing homes. The next section will help you navigate your drainpipes to identify these potential greywater sources. If you are inexperienced with plumbing, this aspect may feel confusing; consider finding a handy friend or plumber to be your reading buddy. Or, read on to develop your greywater sleuthing skills; it can be fun and empowering to uncover the mysteries of the household plumbing system.

Bathroom (Lavatory) Sinks

Since they typically produce such small quantities of greywater, bathroom sinks don't warrant a big investment for a system, though sometimes it's easy to reuse the water. In my house the downstairs sink was easy to divert to irrigate a nearby pomegranate bush and male kiwi vine, whereas the upstairs sink would have been more involved so we just detached the drain to bucket-flush the toilet (we were careful to plug the drain line to prevent sewer gases from entering the bathroom).

Easy options include:

- Combine the sink greywater drain with the shower/bath drain and divert the greywater after both sources have combined.
- Install a diverter valve under the sink and direct water to one or two nearby plants in a tiny branched drain system.
- Alternatively, disconnect the sink drain and collect greywater in a bucket to bucket-flush the toilet. Experiment to find out how much water it takes to flush your toilet: empty the bucket directly into the toilet bowl (not the tank), and the toilet will flush.
- Install a SinkPositive system: a faucet and small sink-bowl that replace the toilet tank lid (see Resources). When you flush the toilet, fill-water flows out the little faucet and through the sink-bowl so you can wash your hands as the toilet tank refills.

There are manufactured greywater systems designed to collect, filter, and disinfect bathroom sink greywater below the sink and then pump it into the toilet tank for flushing. I know a few people who have tried these systems, and each had numerous problems with them. Consider other options first if you plan to reuse sink water.

Kitchen Sinks

Kitchen sinks usually produce a plentiful supply of water that can be diverted from the sink drain inside the house. Kitchen greywater tends to contain food scraps and grease, so it takes more effort to maintain the system than with those for other greywater sources (see Filtering Kitchen Greywater with Mulch Basins on page 16).

Some states consider kitchen water "greywater," while others consider it "blackwater,"

KITCHEN SINK DIVERTER OPTIONS

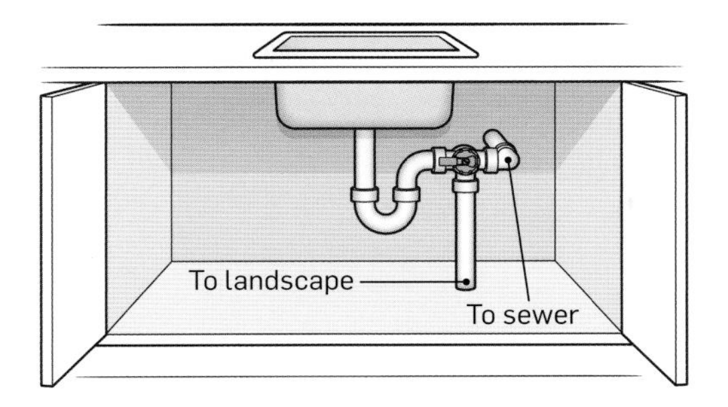

A. Single sink basin with diverter valve

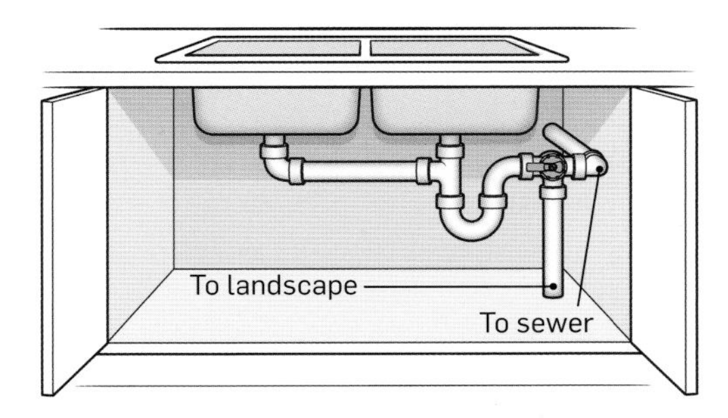

B. Double sink basin with diverter valve

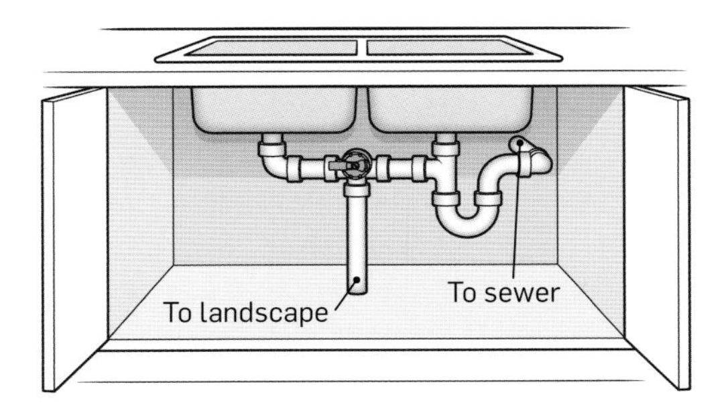

C. Double sink basin with one side of sink connected to greywater system

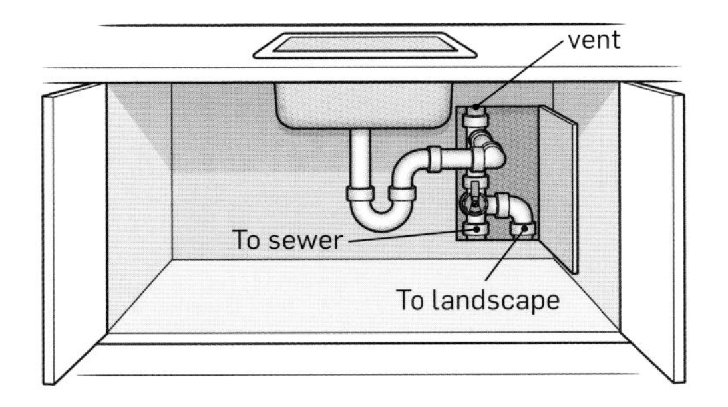

D. Installation of diverter valve after the vent connection

like what comes out of the toilets. If your state doesn't call kitchen water "grey," a legal installation will be more challenging. With determination and an open-minded building department, it's possible to get an experimental permit or use the "alternative materials and methods" section of your state's code.

Divert kitchen water directly below the sink for easy access to the pipes and diverter valve. The greywater pipe needs a route to the landscape, and you can send it below the floor or directly out of the house, depending on your situation and climate. Local code may require the diverter valve be located downstream of the vent connection.

FILTERING KITCHEN GREYWATER WITH MULCH BASINS

The challenge with a kitchen sink greywater system is the gunk. First, make sure you have a fine screen in the sink to catch food BEFORE it goes down the drain. If you have a double sink, consider plumbing just one side to the greywater system and using the other side for the dirtiest, greasiest dishes. Next, think about where the food particles that make it past your screen will end up. Should you have a pre-filter before the irrigation system to catch them? A wood-chip biofilter? Screen? Worm bin? Beware. Any filter will quickly clog, causing a backup or overflow of greywater, plus more maintenance to clean it.

Kitchen greywater creates a unique type of gunk that is merciless on filters. For years I pre-filtered the kitchen water as part of my greywater system. For years I cleaned out disgusting filters: screens, sand, wood-chip filters, wetlands filters. The last straw was the day I came home from a particularly long day at work to see my wood-chip bio-filter had clogged (again) and kitchen sink water was pouring out the top of the filter and running down the side of my house.

Then I found a simple solution: mulch basins (see page 42). These basins are depressions in the soil, filled with wood chips that catch organic matter and infiltrate greywater into the soil. Instead of sending all the water to one basin, divide up the flow to multiple basins. This **branched drain system** requires less frequent maintenance since each basin receives only a portion of the total flow. Organic matter in the greywater decomposes in the basins and is eaten by earthworms.

Mulch basin filtering kitchen sink water. Note that a lid (not shown) covers the outlet.

The first time I dug under my kitchen sink greywater outlets, I was amazed to see the soil beneath the mulch swarming with earthworms. In the past, I'd seen only failed experiments in which the kitchen water flowed directly through a worm bin before the irrigation system. It's hard to keep worms happy when they're doused with water on a daily basis; they leave or die. In the mulch basin system, worms can come and go as they please, eating organic debris from the greywater. Eventually the wood chips will clog and decompose, slowing the infiltration. Maintenance is easy: grab a shovel, dig out the decomposed material, and replace with fresh wood chips.

Other people have had success sending unfiltered kitchen water to large, underground "infiltration galleys" (see page 164 for more details). Using a manufactured grease-trap (or grease interceptor) can also prevent grease from entering the system.

Your Home's Drain, Waste, and Vent System

Your home plumbing system is actually made up of three different systems: the water supply, the drainage/waste system, and the venting system. Drain pipes and vents are connected and commonly referred to as the **drain-waste-vent (DWV) system**. *Water supply pipes* are much smaller in diameter (typically ¾ inch and smaller) than drainpipes (1½ inches and larger) and are full of pressurized water — not a pipe you want to cut into by mistake! *Supply pipes* are commonly made of copper or galvanized steel pipe or CPVC or PEX plastic tubing. *Drainpipes* may be made from cast iron, galvanized steel, copper, or PVC or ABS plastic pipe.

The drain or waste system carries wastewater away from the fixtures to the sewer line or septic system. The venting system is composed of pipes traveling up through the roof, to safely release sewer gases away from your living space and to introduce air into the system. The air prevents a suction effect that can slow the travel of water (think of how holding your finger over the end of a drinking straw "magically" keeps the liquid suspended inside the straw; without venting, the same thing can happen in a drainpipe).

Drainpipes serving sinks, tubs, and other fixtures have **traps**, simple devices to separate the open end of the drain from the sewer end, keeping sewer gases from rising into the living space. The U-shaped section of pipe under your bathroom or kitchen sink is called a **P-trap**. Toilets have traps built into their bases. The bend in a trap retains enough water at all times to seal off the drainpipe; it's self-replenishing and refills every time water flows through the pipe. Allowing air into the system via vents maintains an equilibrium in the pipes so the water in the traps isn't siphoned out. Before homes were plumbed with vents, sewer stench (as well as the occasional explosion due to buildup of sewer gas) was commonplace.

If you look under your bathroom or kitchen sink, you'll see the P-trap between the sink

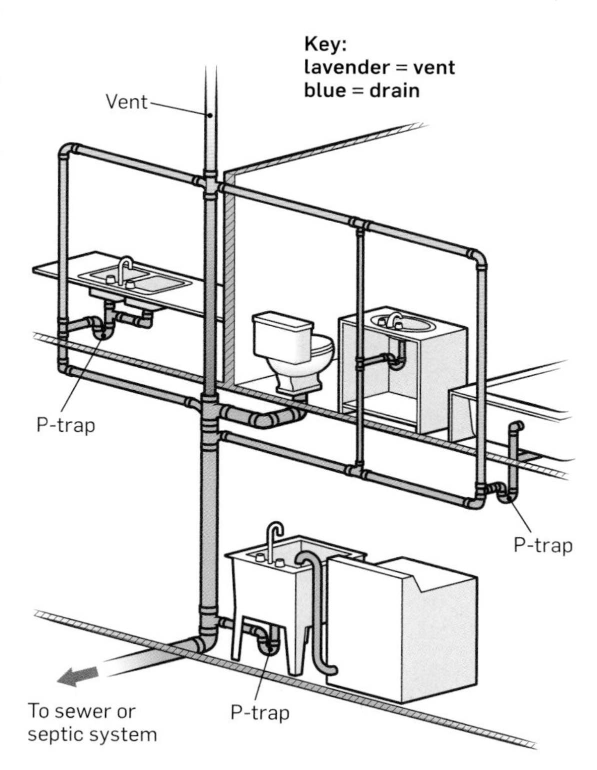

The home drain-waste-vent system. This is a conceptual image and may not represent code-compliant practices.

outlet and a horizontal pipe leading to a hole in the wall. Behind the wall is the vent (going upward) and the drainpipe (sloping downward) connecting to other drainpipes and ultimately to the main drain of the house. Sewer gases rise up and exit the vent above the house, while greywater and blackwater (from the toilets) flow downward to the sewer or septic system. All the drainpipes should have a **cleanout**, a Y-shaped fitting (or a short section of pipe) with a removable cap so you can access the pipes to clean out clogs in the system.

For a well-functioning DWV system it's important that the pipes and vents are large enough to accommodate the flows inside them and that they are sloped properly to avoid blockages. Greywater drainpipes typically are 1½-inch pipe and larger (the outside pipe diameter will be closer to 2 inches), while toilet pipes are 3 inches or larger. Each fixture must be vented, either with its own vent or connected through a secondary vent to the main vent (the uppermost portion of the main drainpipe, sometimes called the "stack"). In general, any reference to pipe size (for example a "2-inch pipe"), refers to the *inside* diameter of the pipe: the *outside* diameter is larger and varies depending on the wall thickness of the pipe material. The outside of a 2-inch copper pipe is smaller than a 2-inch cast iron pipe, even though they have the same inside diameter.

Examining Your Pipes for Diverter Valve Locations

While some older houses (in mild climates) may have exterior drain and vent pipes, most homes have internal piping running inside the wall and floor structures. Drainpipes for greywater systems often are accessed in a basement or crawl space. Unfortunately, houses with slab foundations have their greywater pipes buried in concrete; these pipes are much more difficult to access. If you live in an old house, keep in mind that it may not be properly plumbed. For example, you may find the water from two showers flowing out a drainpipe that is sized for just one bathroom sink. *Always* confirm that you have identified the right drainpipe by running hot water in the fixture you believe it is connected to. The drainpipe will feel warm if your sleuthing is correct.

DIVERTER VALVE BASICS

As you assess pipes for potential greywater use, focus on finding a location for the diverter valve. To install the valve, you'll remove a section of the drainpipe and essentially splice the valve into the pipe. The best location for the valve depends on the source of greywater (sink, shower, etc.) and your home's plumbing configuration. In most situations, the valve should be installed after the vent extends upwards and it *must* be installed *before* the connection to the toilet drain, a 3-inch or larger pipe. You'll need about 7 inches of straight pipe to fit the valve in.

Diverter valve is installed in shower drainpipe after the trap and vent. Note that the backwater valve is installed before the sewer connection.

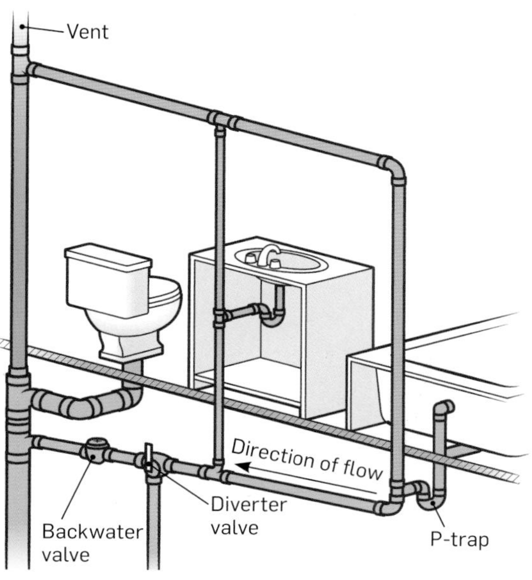

Diverter valve is installed in greywater drainpipe before pipe connects to toilet drain.

WHEN TO CALL IN A PROFESSIONAL PLUMBER

The skill level required to install a diverter valve into existing plumbing depends upon the type and configuration of the existing pipes. If your pipes are made of ABS plastic and there is a large enough section of straight pipe for you to install the diverter valve into, you can most likely do this work yourself. Here are cases when an experienced plumber should install the valve:

→ Your pipes are made from leaded cast iron or old steel, and you don't want leaks to result from the installation.

→ You need to reconfigure the pipes to create space to install the valve.

→ You have a very narrow crawlspace that is hard to work in.

→ You need to install the valve in the wall or floor with an access panel

If you do call upon a plumber, remember they most likely are not trained in greywater. In this situation **be sure you purchase the valve** and adjust it (see page 128) or show them. Keep in mind that many of the fundamental principles of greywater irrigation directly contradict the training and experience of plumbers. Be sure to find a plumber who is open to learning something new!

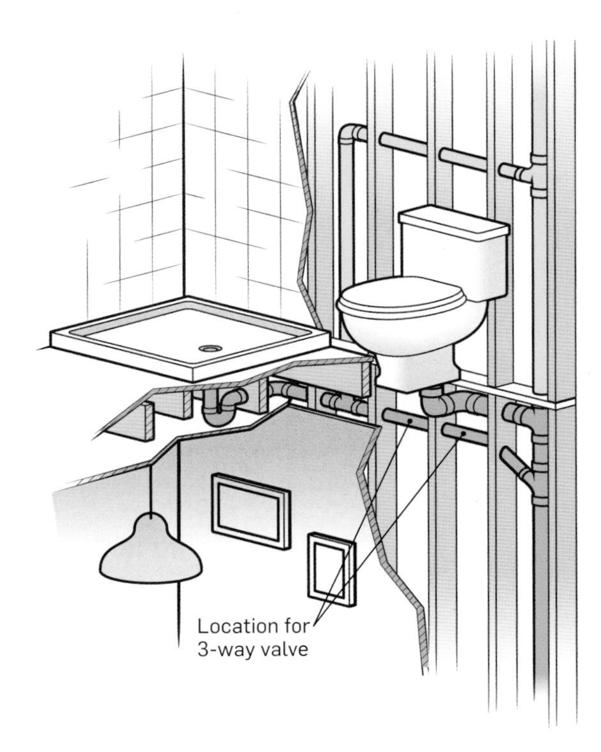

INTERNAL PLUMBING. *Shower and toilet drains combine inside the wall of a second-story bathroom. Note potential location for diverter valve.*

If there is not a large enough section of straight pipe to fit the diverter into, you'll need to reconfigure the plumbing to create room. If you're new to plumbing, this is an excellent time to get assistance from an experienced plumber.

HOUSES WITH INTERNAL PLUMBING

To identify first-floor shower pipes, go into the crawl space or basement and look for a P-trap. You may even see part of the bathtub above the P-trap. To make sure you have the correct pipe, run hot water down the drain of the tub or shower and feel which pipe heats up. Second-floor shower drains are commonly combined with the toilet drain before dropping down to the crawl space or basement in a 3-inch pipe — but not always. Check for 2-inch pipes below an upper-story bathroom; run hot water in the shower and feel if the pipe heats up.

The DWV system of some homes is plumbed inside the walls, with no convenient crawl space or basement to access the drains. During a remodel, or by removing a section of the wall, a diverter valve can be installed with an access hatch, to direct greywater to the landscape. In the image at left, the shower drain is inside the wall when it combines with the toilet drain.

STAY SAFE IN THE CRAWL SPACE. The crawl space may not be the safest place, especially if your house is old. Before crawling under, check for potential hazards lurking below, such as mold, damaged asbestos pipes, standing water or wet areas which could cause electric shock if the wiring is improperly installed, spiders, or animal presence (you don't want to encounter an angry raccoon when you're crawling on your belly with no quick escape route!). After you have identified potential hazards, take the necessary precautions to stay safe (gloves, respiratory masks, eye protection, thick clothing covering all your skin, flashlight, turn off the power, etc.). If wild animals nest under your house, wait till they've left, then board up their entrance before you begin looking for potential greywater sources.

SOLUTIONS FOR GREYWATER PIPES BURIED IN THE WALL/FLOOR/CEILING

It takes more effort (and cost) to access greywater pipes inside the wall or floor. Often this requires cutting out a section out of the floor/wall/ceiling finish to access the pipe and install the valve, as well as making an inspection hatch or access door so you can get to the valve to control the flow of greywater. An alternative for a bathtub drain is to elevate the tub on a platform to create room for the valve in the drain before it goes into the floor; however, this is a big project and best done during a remodel.

If your shower pipes are buried in a concrete slab foundation, you have a few options. During a remodel, keep the shower pipe separate from the toilet until they're outside the house, where you can install a valve in a subsurface access box. Or, elevate the tub and install the valve in the drain before the pipes enter the slab.

If you can't feasibly access the pipes, consider using buckets, siphons, or installing an alternate shower that is greywater accessible.

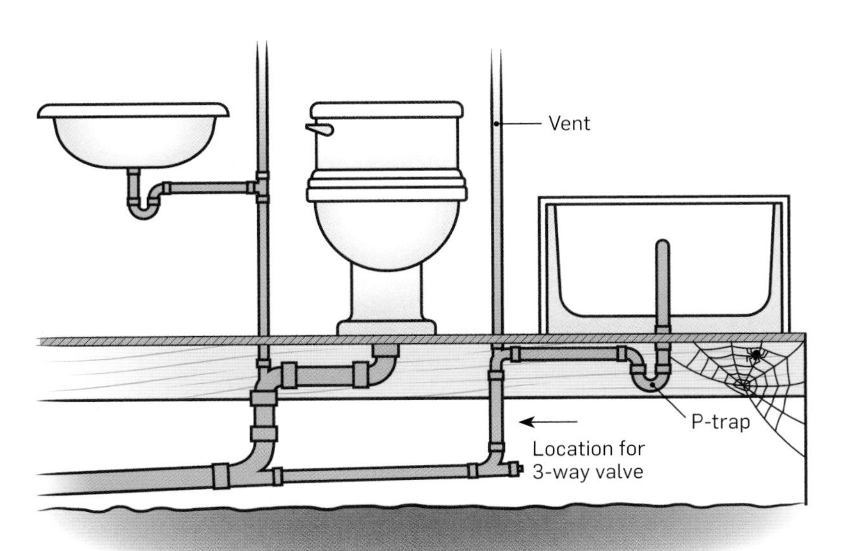

INTERNAL PLUMBING. *Plumbing is accessible in the crawl space under this first-floor bathroom. Note potential location for diverter valve.*

GREYWATER-READY BUILDING CONSTRUCTION

If you construct a home with a future greywater system in mind, it will be less costly and easier to install the future system. Architects, designers, builders, and developers can plan these "greywater-ready" buildings using a few basic design principals. Local or statewide ordinances can require this practice.

1. Estimate how much greywater will be generated and how much of the landscape can be irrigated. Plan to access to enough of the greywater sources so the landscape can be irrigated entirely with greywater.

2. Create greywater accessibility points so that the greywater drains are accessible and able to be diverted into a greywater irrigation system (top image, this page). If greywater drains can't be accessed below the building (in a crawl space or basement), create an access panel in the wall (top image, next page) or a subsurface access point in the landscape (bottom images, next page) before greywater combines with any toilet water.

3. Keep the greywater accessibility location as high in elevation as possible to allow for gravity-fed systems, and locate it so the greywater can access an irrigated portion of the landscape.

4. If the future diverter valve location won't be easy to access for manual turning of the valve, make sure an electrical outlet is located nearby so an electronic actuator can be connected to the diverter valve for remote switching.

Greywater-Ready Building Options

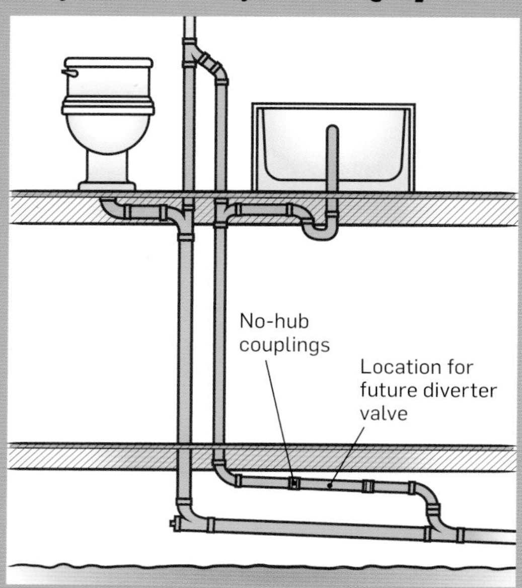

Greywater pipe runs into crawl space separately from blackwater pipe. Two no-hub couplings make it easy to remove a section of pipe and install the future diverter valve.

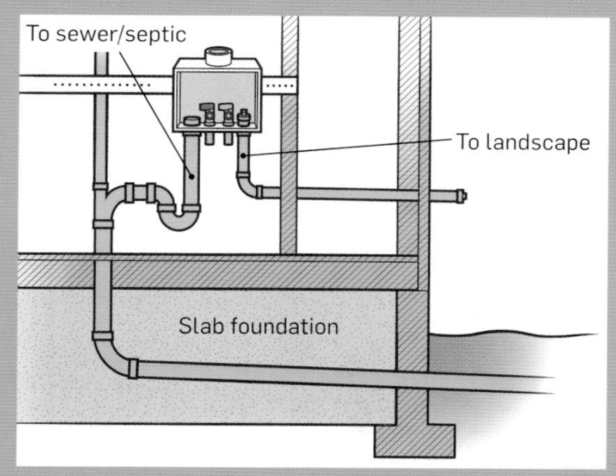

Install a dual-drain washer box. Plumb one drain conventionally to the sewer/septic, and outfit the other drain with a pipe to the outside of the house. A 1" flexible tube can be snaked through this pipe for a future L2L system.

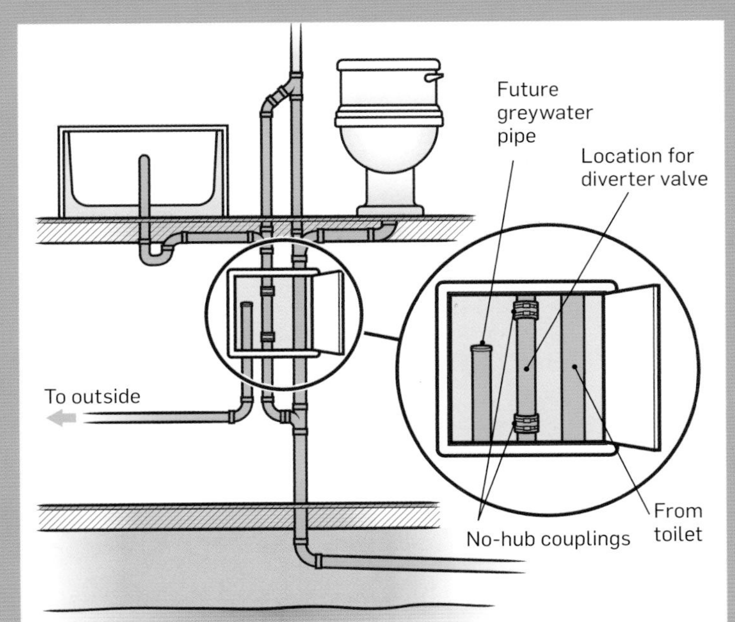

Before closing up the walls, install a greywater pipe to the outside of the building with an access panel where the diverter valve will be located. The diverter valve will direct greywater either into the pipe to the landscape or into the existing pipe to the sewer.

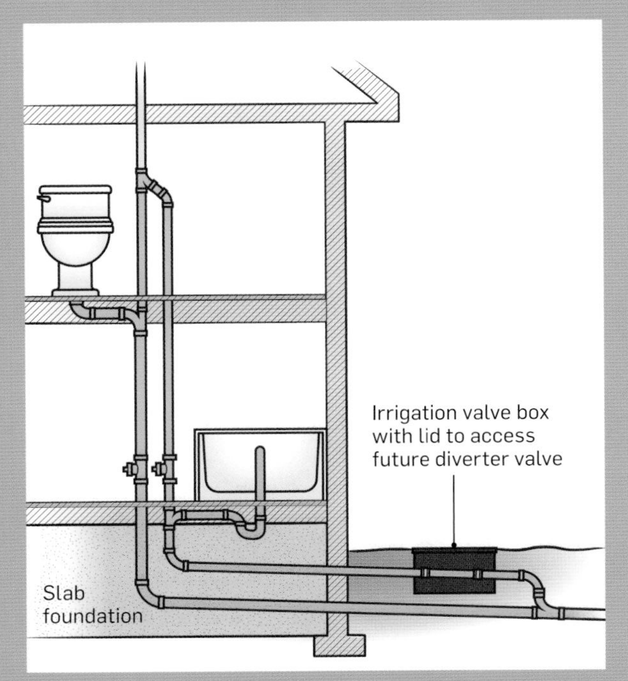

Keep greywater pipe separate from the toilet pipe until outside the building. Locate an irrigation valve box above the location for the diverter valve for future access.

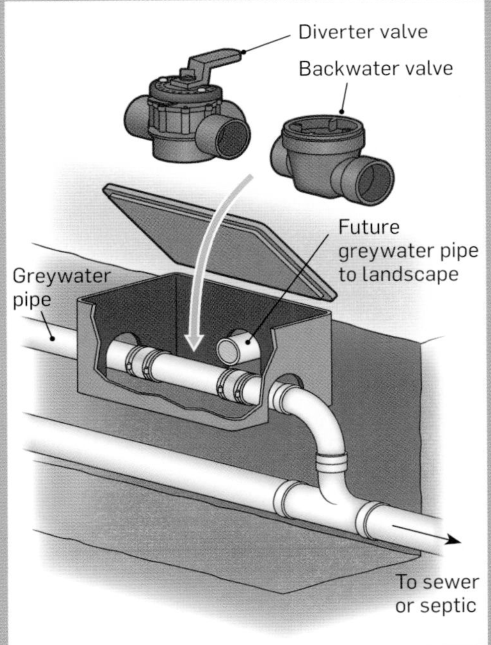

Close-up of inside valve box and where to install future diverter valve

HOUSES WITH EXTERNAL PLUMBING

If you see pipes running up and down the outside of a wall near a bathroom, these are likely to be exterior vent and drain pipes. Exterior pipes make it easy to identify and access greywater sources, and you can divert the water without slithering under the house.

Since exterior pipes from a bathroom may contain any combination of sink, shower, and toilet, take care to properly identify what fixtures flow into the pipes. For example, a 2-inch or larger pipe that you believe is the shower drain may also contain the sink water. Test the pipes by running hot water from each fixture. Also, note where the fixture's drainpipe enters the exterior pipe. Sometimes this isn't obvious. The tee fitting may be buried in stucco or covered with siding. If you accidentally cut into the pipe too high, you will be in the vent where no greywater flows.

Notice in the image at right, the tub drainpipe combines with the toilet drainpipe before exiting the wall. To divert the shower water, a diverter valve would be located under the floor of the bathroom. The tub drainpipe in the image on page 25 exits the house separately from the toilet drainpipe. The diverter valve could be located anywhere along the drainpipe and could be placed to collect water from the washing machine or sink drains as well. Locate the valve to collect from the desired fixtures, and in the most accessible location.

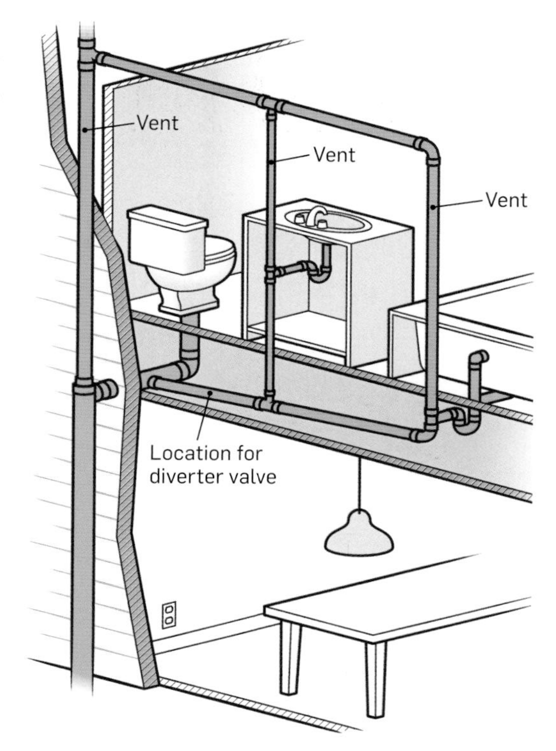

EXTERNAL PLUMBING. *Shower and toilet drains combine in floor. Diverter valve must be located in the floor/ceiling.*

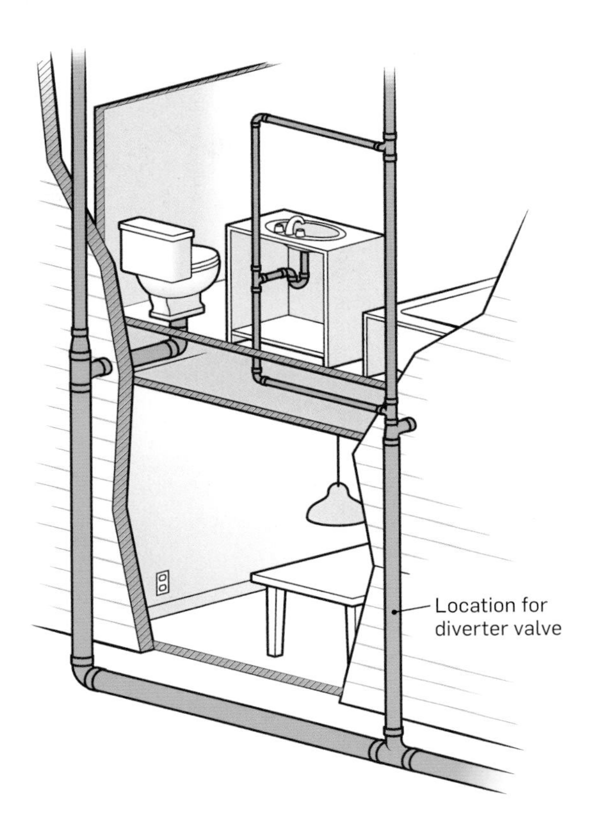

EXTERNAL PLUMBING. *Shower/sink and toilet drain pipes exit the wall separately. The diverter valve can be located outside.*

Undesirable Greywater Sources

Not all sources of household water are appropriate to reuse in a backyard system. Don't use wastewater from:

- Dishwasher drain water. Most dishwasher detergents are high in salt content. If you don't have a sewer/septic option, and can't find a plant-friendly detergent, send the dishwasher water to irrigate salt-tolerant plants.
- Water drained from any sinks that receive chemicals, such as in darkrooms.
- Softened water from a sodium-based water softener, which contains salts that are harmful to plants. Use a potassium-based softener, physical water conditioners, or magnetic softener instead if you want to reuse the greywater or, if possible, route unsoftened water to the washing machine.
- Water softener backwash; also high in salts.
- Toilet wastewater. *Never* put toilet wastewater into a greywater system.

Summary of Greywater Sources

Here's a quick rundown of the major greywater sources, including the advantages and disadvantages of each.

Clothes Washer

ADVANTAGES: The washer's internal pump and flexible drain hose make accessing water easy and allows you to tap into the system without altering the existing plumbing. It's usually easy to send a new greywater pipe outside, either through a crawl space or exiting through a wall. It's also easy to turn off the greywater system if there is too much water or you want to use bleach in a wash load.

DISADVANTAGES: The machine's internal pump is not strong enough to pump water uphill or across long distances, and the system is unable to distribute water over a large area.

NOTES: If the laundry room is in an interior room and there is no crawl space or basement, sending a pipe outside can be relatively difficult or impractical. Many states do not require a permit for a washing machine system that does not alter the existing household plumbing.

Showers and Baths

ADVANTAGES: Often the largest and cleanest source of greywater.

DISADVANTAGES: Installing the diverter valve can be tricky, depending on the configuration of the existing plumbing. There must be room to access (and, if necessary, alter) the drainage plumbing, either in a basement or crawl space or below the floor.

NOTES: If you have old plumbing or a tricky install, or if you are not experienced with plumbing, consider hiring someone to install the diverter valve. Showers on a concrete slab or in a second-story bathroom pose challenges for installing a diverter. Consider an outdoor shower if you are unable to capture from the indoor shower(s).

Bathroom Sink

ADVANTAGES: Easy access to greywater from under the sink. Often easy to combine the sink drain with the shower drain under the bathroom.

DISADVANTAGES: Small quantity of water with relatively high concentration of soap, toothpaste, and other personal care products.

Kitchen Sink

ADVANTAGES: Easy access to water under the sink. Organic matter from food particles creates compost and is good for gardens.

DISADVANTAGES: Because there is additional organic matter in kitchen greywater the systems are more prone to clogging.

NOTES: Some states don't consider kitchen greywater "grey"; they classify it with the toilet wastewater and don't allow legal reuse. If the sink is a double sink, consider hooking up just one side to the system (the side without the garbage disposal, if there is one).

Other Potential Sources

These sources are not technically considered greywater but often can be reused for irrigation.

REVERSE OSMOSIS (RO) WATER-FILTER DISCHARGE: Depending on the model and age of the system, RO filters will discharge 4 to 20 gallons of backwash water for every gallon of filtered water it produces. This is good-quality irrigation water and can either be combined with other greywater sources or used in a stand-alone system.

AIR CONDITIONER CONDENSATE: High-quality irrigation water that can used for irrigation, either combined with other greywater sources or used as a stand-alone system. More water is produced in warm, humid climates than in warm dry climates, with a typical home air conditioner unit producing between 5 to 20 gallons a day. Window AC units produce just a gallon or two a day.

EVAPORATIVE COOLER (SWAMP COOLER): Bleed-off water is another potential source of water, with discharge rates of 1 to 5 gallons per hour, though this water is high in salts and should only be used to irrigate salt-tolerant vegetation.

CHAPTER 3

Estimate Your Greywater Flows

Next you'll need to find out how much greywater your home produces. This depends on the number of people in your house, your water use habits, and your fixtures and appliances.

On average, each American produces between 20 and 40 gallons per day of greywater. It's important to get the most accurate estimate you can before designing your system to avoid over- or under-watering your plants. This chapter leads you through the basic calculations for both personal greywater production and estimates to satisfy local code requirements. I'll also discuss effects of diverting all greywater sources from a sewer or septic system.

IN THIS CHAPTER:

Code Estimates vs. Personal Calculations

If your project involves a permit (see chapter 7), you may be required to use a method specific to your state or local code (see Sample "Code-Type" Estimate for Greywater Production on page 35). Yet even if you use a code method, it's a very good idea to also use the method described on page 30, because it's specific to your habits and therefore gives you the best estimate for how much irrigation water is available.

State codes tend to overestimate greywater flows. This ensures the system is large enough to handle volume increases — not a bad idea if your in-laws decide to move in. The downside is that when the home produces less greywater than estimated, the system will be overbuilt, will cost more, and may not irrigate properly. In places with tiny yards, this could prohibit a legal installation due to an insufficient landscape area to accommodate the oversized system. If your system does not involve a permit, you're better off avoiding the code method since it's not very accurate.

To estimate how much greywater your household produces, you'll need a few pieces of information:

1. The flow rates of your fixtures.
2. The bathing and laundry habits of everyone who lives in your house (number and length of showers per week, number of baths per week, and loads of laundry washed).
3. Information regarding future changes in occupants or habits. For example, are you about to have a baby? If so, laundry use will increase. Are your kids leaving for college soon? If so, shower usage will decrease.

Design your system using the most accurate estimate for the foreseeable future, and consider ways to adapt for future changes. For example, a greywater shower system may be a perfect fit with kids leaving for college and new trees needing irrigation: the young trees will benefit from the extra greywater while the kids are still at home, and once the trees are established they can survive on less after the kids leave. If your usage is going to increase, think about how you might expand the system or create another zone to utilize more water.

Use personal calculations to estimate how much greywater your home produces so that you deliver the right amount of water to your plants.

Calculating Weekly Greywater Flows

Use the following formulas to calculate your weekly greywater output. Also see Finding the Flow Rates of Different Fixtures (page 32) to determine actual flow rates of your fixtures.

WASHING MACHINE

WEEKLY PRODUCTION	=	**gallons/load** *(the rating of your washing machine)*	×	**loads per week**	=	**gallons per week**

Also calculate the maximum daily flow from your washing machine so you can appropriately size the landscape area to avoid pooling or runoff of greywater. For example, if you do all the laundry on Saturdays, you'll need a larger infiltration area than if you spread out the laundry over the week.

MAXIMUM DAILY FLOW	=	**gallons/load** *(the rating of your machine)*	×	**loads on a typical laundry day**	=	**maximum gallons per day**

SHOWERS

WEEKLY PRODUCTION	=	**gallons per minute** *(or gpm, which is the flow rate of your showerhead)*	×	**minutes per shower**	×	**number of showers per week** *(if people in the house take different lengths of showers, alter this formula accordingly)*	=	**gallons per week**

(Repeat this formula as needed for each person in the home.) Typical shower flow rates are 1.5 gpm to 2.5 gpm.

BATHS

MAXIMUM DAILY FLOW	**gallons per bath** *(typically between 30 and 45 gallons)*	×	**baths per week**	=	**gallons per week**

BATHROOM SINKS

MAXIMUM DAILY FLOW	gallons per minute	×	minutes per use	×	uses per day	=	gallons per day	×	seven *(days in week)*	=	gallons per week

Bathroom sinks are typically 2 gpm to 2.2 gpm, or 2 gallons per person per day (gpcd).

KITCHEN SINKS

MAXIMUM DAILY FLOW	gallons per minute	×	minutes per use	×	uses per day	=	gallons per day	×	seven *(days in week)*	=	gallons per week

Kitchen sinks are typically 2 gpm to 2.2 gpm or 3 to 8 gallons per person per day (gpcd).

Compare Your Results to a Water Bill

After calculating your weekly water use, compare the results to a winter water bill, a time of year when you do not irrigate. Greywater is typically 75 percent of total indoor use. For example, if you estimated 80 gallons of greywater per day, and your December water bill shows 110 gallons per day, your estimate is close to 75 percent (110 gallons × 0.75 = 82.5 gallons). If your number is not close to 75 percent of the winter bill, go back and check your results. If you feel confident your usage estimates are accurate and your estimate is much lower than what the bill shows, you may have a water leak.

Baths typically use more water than showers.

FINDING THE FLOW RATES
of Different Fixtures

Most fixtures have their flow rates printed on them in small type, though their actual flow rate may be different due to variations in water pressure. Finding the actual flow rate is easy, especially if you have a helper. If you find that your showerhead uses more than 2.5 gpm, replace it!

MATERIALS

- **Large container, such as a 5-gallon bucket or large pot**
- **Timer that measures seconds**
- **Container that measures gallons (such as a gallon-size milk jug)**

Calculating Showerhead or Faucet Flow

1. Turn on the shower or sink faucet full-blast.
2. Using the timer, fill the large container for exactly 1 minute.
3. Pour the water into the 1-gallon container, emptying and refilling the container as needed and noting how much water was in the bucket. Round the total to the nearest ¼ gallon. This is your fixture's flow rate in gallons per minute (gpm).

Calculating Washing Machine Gallons per Load

Method 1: Get a large container (or several smaller ones) and fill it with the drain hose of the machine as you wash a load of laundry. You must collect the drained water from both the wash and rinse cycles. If you have an older top loader, be prepared to collect up to 50 gallons of water! This is the best and most accurate method.

Method 2: Call a retailer or manufacturer that carries the same brand machine you have and ask them the typical gallons per load of your model machine. This information is not easy to find on the Internet. Keep in mind that many machines now adjust water use based on the size of the load, which makes it harder to determine a "typical" load.

Sample Calculation of Greywater Irrigation Potential

Here's an example of how four people in a household can calculate their total gallons-per-week greywater output, and thus their irrigation potential.

SHOWER

Two people take one 3-minute shower every day. Two people take 8-minute showers four times a week. The showerhead's flow rate is 2.0 gpm.

2 gpm *(showerhead flow rate)*	×	3 *(minutes)*	×	7 *(showers per week)*	×	2 *(people)*	=	**84 gallons**
2 gpm *(showerhead flow rate)*	×	8 *(minutes)*	×	4 *(showers per week)*	×	2 *(people)*	=	**128 gallons**

TOTAL GALLONS PER WEEK = 84 + 128 = 212 gallons per week

BATHROOM SINK

1.5 gpm × 1 minute per person per day

GALLONS PER WEEK = 1.5 gallons × 4 *(people)* **× 7** *(days in a week)* **= 42 gallons**

WASHING MACHINE

The entire household does five loads of laundry each week. Their machine uses 20 gallons per load.

GALLONS PER WEEK = 5 *(loads per week)* **× 20** *(gallons per load)* **= 100 gallons**

KITCHEN SINK

Flow rate of the sink is 2 gpm. The household estimated they used the sink for 4 minutes on breakfast dishes, and used dishpans for the dinner dishes (2 gallons to wash and 3 gallons to rinse).

GALLONS PER WEEK = (2 gpm × 4 *(minutes)* **+ 5 gallons** *(tubs)* **) × 7** *(days in a week)* **= 91 gallons**

Total Weekly Greywater Flow: 445 gallons

RECORD YOUR HOME'S GREYWATER OUTPUT

Fixture	Flow rate (gpm) or total gallons	Typical usage (for reference)	Minutes per use	Uses per week	= Weekly greywater (gallons)
WASHING MACHINE	(gallons/load)	Top-loader 40–50 gallons/load Top-efficient 20–30 gallons/load Front loader 12–25 gallons/load	N/A		
SHOWER		Low flow = 10 gallons/person/shower High flow = 20 gallons/person/shower			
BATH	(gallons/bath)	30–45 gallons per bath	N/A		
BATHROOM SINK		1–5 gallons/person/day			
KITCHEN SINK		3–8 gallons/person/day			
OTHER:					
OTHER:					

SAMPLE "CODE-TYPE" ESTIMATE FOR GREYWATER PRODUCTION

Note: This example uses the California Plumbing Code (2015 version); other states may use different numbers.

STEP 1: Determine the number of people in the home based on the number of bedrooms. Count two occupants for the first bedroom and one occupant in each additional bedroom. It doesn't matter how many people actually live in the house.

STEP 2: Calculate daily water use. Showers, bathtubs, and bathroom (lavatory) sinks combined produce 25 gallons per day (gpd) per occupant. Washing machines produce 15 gpd per occupant.

STEP 3: Multiply the number of occupants (based on Step 1's result) by the water use (per occupant) to determine the total estimated greywater produced daily:

Number of occupants × daily water use = total daily estimated greywater flow

For example, a three-bedroom home would produce:

→ **Number of occupants:** 4 (2 in first bedroom, 1 in each of the other bedrooms)

→ **Shower and sink water use:** 25 gpd × 4 occupants = 100 gallons per day or 700 gallons per week

→ **Washing machine greywater:** 15 gpd × 4 people = 60 gallons per day or 420 gallons per week

Total greywater = 160 gallons per day or 1,120 gallons per week

If two people lived in this home they would probably produce much less greywater than 160 gallons per day (1,120 gallons per week), especially using water-efficient fixtures. To comply with a code using this method the occupants would size the infiltration area of the system to accommodate 160 gallons per day of greywater. Some codes allow alternative estimation methods, such as reduced estimates in homes with water-efficient fixtures.

DIVERTING 100 PERCENT OF YOUR GREYWATER AWAY FROM THE SEWER OR SEPTIC DRAIN

A home's sewer or septic drain needs enough wastewater flowing in the underground pipe to get the poop and toilet paper (the solids) from the toilet to the city's main sewer line (or a private septic tank) to prevent clogs. In a properly functioning home system, each toilet flush carries the solids all the way to the main line. Unfortunately, many homes' sewer drainpipes are not sloped properly and/or have root intrusion or other issues that affect their performance, and greywater (from showers, sinks, etc.) picks up the slack, helping to carry the solids through the pipe.

Greywater "Backup": To Have or Not to Have

My house diverted all greywater to the landscape, and only one ultra-low flow toilet was connected to the sewer. For a few years there were no problems. Then the house's sewer drain clogged. A plumber unclogged it and said it clogged because we didn't have enough water flowing through the system. We ignored him. It clogged again. We redirected one shower to the sewer. It clogged again. Finally, a company sent a camera down the sewer line and discovered the old clay sewer pipe was collapsed and missing in parts, and had roots inside it.

Our problems were not related to low greywater flows but rather resulted from a 100-year-old broken clay sewer pipe. After our new, properly sloped sewer line was installed we diverted all the greywater to the yard problem-free (about 7 years have passed now). The lesson: If you divert all your greywater to the yard (and you haven't abandoned the sewer completely with composting toilets), be prepared to troubleshoot sewer issues that come up, knowing that regardless of the true source of the problem your greywater system will get blamed.

The Bigger Picture

What would happen to the entire municipal sewer system if lots of people reused greywater? Massive sewage clogs? Probably not. In the *Study of the Effects of On-Site Greywater Reuse on Municipal Sewer Systems* (2011; see Resources) researchers used hydrologic models of the sewer system to identify levels of flows and pollutant loads throughout the day. They found the sewer system operated with large fluctuations in flows, with two major peaks occurring each day (morning and evening). Reusing greywater would have the largest impact on the system during these peak flow times (imagine lots of morning bathing water going into the yard instead of the sewer). Widespread reuse of greywater would reduce the peak flows but would have almost no impact on the system during the current lowest-flow times of the day (when toilet flushing is the major source of water in the system).

Because the sewer system currently operates at these low-flow times, and these flow rates would not be affected, researchers concluded that widespread reuse of greywater should not increase blockages. In fact, they hypothesized that sewer systems could experience positive effects from greywater reuse with an increase in capacity.

Septic Considerations

A septic system also needs sufficient wastewater to function properly. Is it okay to remove all greywater from the septic? Some people say no, that it will cause septic failures. Yet many homes have done this without problems. In fact, diverting greywater from the septic helps solve the most common problem with failing systems: too much water. According to ecological wastewater designer Bill Wilson, "By tapping out the greywater, all you are doing is greatly extending the retention time. This makes for better primary treatment, not worse." One system Wilson designed for a winery tasting room sent water only from a very efficient dishwasher, foam flush toilets (3 ounces of water per flush), and waterless urinals, all discharging directly into a properly designed septic tank — problem-free. A properly designed system has an effluent filter and sufficient distance between the inlet and outlet to assure good separation of liquids and solids, among other important details (see Resources for more information). If you have, or are planning to install, an advanced on-site wastewater system (not the traditional septic tank with leach field) be sure to talk to your system designer about how potential changes to your flows may affect the system before diverting multiple greywater sources to the landscape.

Note that there is an alternative to reusing greywater for homes on septics: reusing septic effluent directly from the tank; see pages 174 to 177 for more information.

CHAPTER 4

Soils and Mulch Basins

The next step in designing your greywater system is examining the soils found on your site.

In this chapter you'll learn how to perform a simple soil test to identify your soil type and to determine how quickly water soaks into the ground. You'll use this information to size your mulch basins so that greywater has enough area to soak into the ground, rather than pooling or running off. I'll also discuss important design considerations that ensure that your system is protective of groundwater and drinking water wells.

IN THIS CHAPTER:

- → Soil Structure and Type
- → Identify Your Soil Type with a Soil Ribbon Test
- → Conduct a Simple Infiltration Test
- → Mulch Basins
- → Protect Groundwater and Drinking Water Wells

Soil Structure and Type

Healthy plants require healthy soils. The soil in your yard is composed of different types and sizes of particles. If you crush a handful of dry soil in your hand, you may notice mineral particles (from broken-down rocks) of different sizes — some very tiny and others, such as sand grains, larger and visible to the naked eye. The soil texture, the size of the soil's mineral particles, is static: you cannot change it.

In your handful of soil you'll also notice organic particles from decomposing leaves, sticks, roots, and other plant material. These mineral and organic particles are held together to form clumps of soil called **aggregates**. In between the clumps are air spaces, which allow oxygen and water to move through the soil. A suitable soil structure, with enough air spaces between soil clumps, is important for plant health.

What holds these soil particles together? Soil microbes. These little creatures not only break down organic matter and make nutrients available to plants, but they also excrete a gluelike substance that holds soil particles together, building soil structure. Soils that have poor structure, either from compaction or lack of organic matter, can be improved by adding compost and planting deep-rooting plants to increase aeration in the soil.

Get to Know Your Soil

It is important to know your soil type, or **soil texture**, when designing a greywater irrigation system. The soil type affects how quickly or slowly water soaks into the soil.

Soils are comprised of clay, sand, and silt particles. Clay particles are the smallest and have the least air space between them. Sand particles are the largest with the most air space. Water moves more slowly through clay soils than sandy soils; greywater irrigation in clay soils requires a larger infiltration area than in sandy soils because it will take longer for the water to soak through clay. If the infiltration area is too small, the soil could become saturated, resulting in greywater pooling on the surface before all the water can seep in.

You can identify your soil type with a simple "ribbon" test. However, be aware that some local codes require a specific method to meet permit requirements, such as a laboratory analysis (see Resources).

IDENTIFY YOUR SOIL TYPE

with a Soil Ribbon Test

Soil type can be highly localized, meaning a home landscape can have different types in different areas. For best results, test your soil directly in each area you plan to irrigate with greywater and below any amended soil; for example, if you've added a layer of compost over the native soil. In most cases, test to a depth of 8 to 12 inches, as this is how deep your mulch basins (page 42) may be.

MATERIALS

- Trowel
- Water
- Ruler

1. **USE THE TROWEL TO LOOSEN THE SOIL** and take a walnut-sized sample into your hand. Remove any rocks or roots from the pile, adding more soil if needed.
2. **MOISTEN THE SOIL SLOWLY,** kneading as you add water until it reaches the consistency of bread dough, not a slurry. If the sample gets too wet or dry, adjust with more soil or water. Mix the paste around on your palm and notice whether the soil is mostly sandy (which indicates sandy soil), has a fine gritty feel (which means it has silt), or is very smooth (which means it is rich in clay). Keep mixing and kneading the soil for a minute or two until it is uniform and any lumps of clay have been thoroughly wetted. You may need to add a little more water to it.
3. **MAKE A BALL OUT OF THE SOIL,** then try to form a ribbon: place the ball in your hand between your thumb and forefinger, gently squeeze the soil, and push it upwards into a ribbon, extending the ribbon as long as it will go before breaking from its weight. Don't try and mold the soil into a ribbon by rolling it in your palms. **Note:** If the soil will not make a ball, it is **sand**; if it makes a ball but not a ribbon, it is **loamy sand**.
4. **IDENTIFY YOUR SOIL TYPE** based on the length of the ribbon:
   - If the ribbon breaks into pieces that are less than 1" (2.5 cm) long, you have some kind of loam. If you can feel many sand grains, it is a **sandy loam**; if not, it is **loam**.
   - If the ribbon is more than 1", perhaps even 2", you have a **clay loam**. If it is sandy-feeling, it is a **sandy clay loam**.
   - If the ribbon is more than 2" long, it is **clay**. It will probably be shiny when wet. If you can feel many sand particles, it is **sandy clay**.

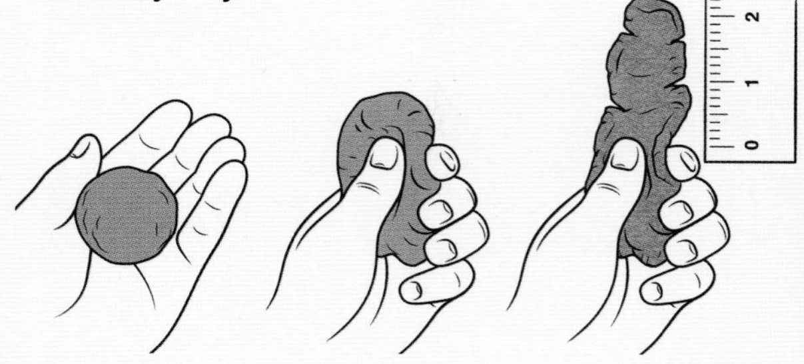

Soil ribbon test

DETERMINE HOW WATER FLOWS THROUGH YOUR SOIL

with an Infiltration Test

How does water flow through your soil? If you're directing greywater to a planted area you currently irrigate you probably already know that water infiltrates there, so this test may be unnecessary. But if you're new to the landscape or are directing greywater to a previously unlandscaped area, conduct a simple infiltration test to make sure the site is suitable for greywater irrigation. Surprises may await you a few feet underground. For example, you might discover that someone has buried a concrete slab underground, preventing water from soaking through; thus, a bad location for greywater.

Sending greywater to a poorly draining area isn't good for the plants or the irrigation system design. Plants don't like soggy roots (unless they're wetland plants), and a slow infiltration rate leads to pooling or runoff of greywater. If you can't direct greywater elsewhere, focus on improving the infiltration of the soil; add compost, use large mulch basins (see page 42) to soak up greywater and release it slowly, and plant deep-rooting plants to open the soil.

Infiltration test

MATERIALS

- **Shovel**
- **Marked stick**
- **Source of water**

1. **DIG A HOLE,** about 12" deep, and as wide as your shovel, in the area you plan to infiltrate greywater. Insert a ruler or stick marked with inches into the hole.

2. **FILL THE HOLE WITH WATER.** Let the water soak into the soil, then repeat. The goal is to saturate the surrounding area, and this usually takes at least three fillings (or conduct the test after a rain).

3. **FILL THE HOLE WITH WATER AND MEASURE THE LEVEL** with the marked stick. Record how long it takes for the water level to drop several inches. If water drops 1" per hour or faster, you have sufficient drainage for irrigating the area with greywater. If it takes longer than 2 hours for the water level to drop 1", or if all the water in the hole doesn't drain all day, this is not a good location for greywater irrigation.

Mulch Basins

A mulch basin is a sunken area in the landscape filled with mulch (wood chips, straw, or other organic material). It's designed to absorb and infiltrate greywater into the ground and prevent it from pooling or running off. Most of the greywater systems described in this book, except drip irrigation systems, use mulch basins. Any food particles and organic matter stick to the mulch and decompose. The size and type of mulch used depends on what's available locally, as well as how long you want it to last. Larger wood chips decompose more slowly than finely shredded material and won't need replacing as often. Typically, large wood chips will last 1 to 3 years.

Sizing Mulch Basins

Suppose you've worked out how much greywater you have, and you decide to irrigate 10 trees. Will your 10 mulch basins be large enough for all the water? To calculate the necessary size of the infiltration area, you need to know the soil type and how many gallons per day of greywater will enter the system. As an example, if you have clay soils, you'll need about 1 square foot of infiltration area per gallon of greywater. The Soil Types and Infiltration Area chart opposite lists basic area requirements for various soil types.

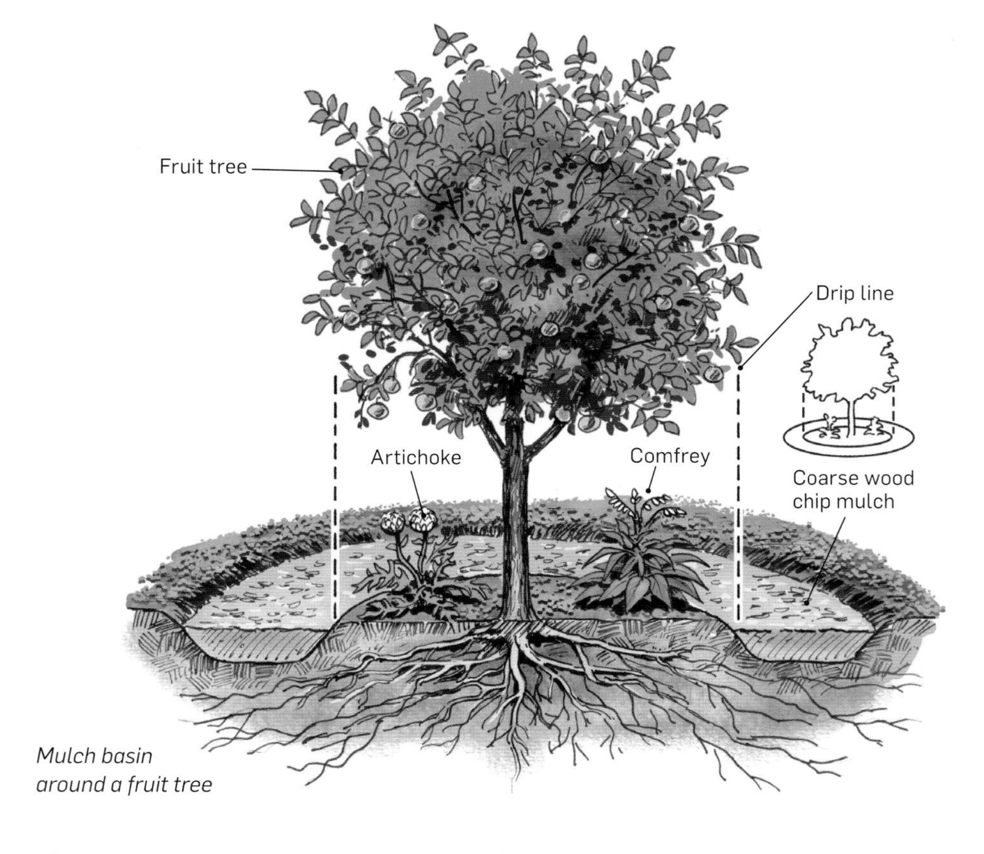

Mulch basin around a fruit tree

SOIL TYPES AND INFILTRATION AREA

Soil type	Area needed to infiltrate each gallon of greywater (per day)
Sand	0.25 square foot
Sandy loam	0.4 square foot
Loam	0.5 square foot
Sandy or loamy clay	0.6 square foot
Clay	1 square foot

Multiply the gallons per day of greywater your home generates by the number corresponding to your soil type. This square footage represents the minimum size of the total mulch basin area your system requires. For example, if a home produces 60 gpd of greywater and the irrigation site has clay soil, the minimum infiltration area required is 60 square feet (60 × 1 = 60 square feet). Since the system irrigates multiple plants, the 60 square feet of infiltration area can be spread out across the yard. For example, to irrigate 10 trees, each would need 6 square feet of infiltration area.

Keep in mind this calculation doesn't consider the plant water needs (see How Much Water Do My Plants Want? on page 49). Also note that systems that use drip irrigation won't use this calculation because the water is spread out over a much larger area; therefore, ponding and runoff are rarely problematic.

SIZE BASINS FOR THE FUTURE

Is your household size increasing? If so, size your infiltration area large enough to accommodate increased flows. Will you be selling your house? If so, it's a good idea to size the system based on the number of bedrooms since you won't know how many people may move into the home (see Sample "Code-Type" Estimate for Greywater Production on page 35).

Protect Groundwater and Drinking Water Wells

Greywater has the potential to contaminate a shallow groundwater table, and in many areas groundwater is used as a local water supply. To prevent this, greywater should infiltrate through at least 3 feet of soil before reaching a water table. Not sure how deep the groundwater is where you live? Get out the shovel and dig a 3-foot-deep hole at the lowest point of your proposed system (see Call Before You Dig! on page 65). If there is no visible water seeping into the hole, then the area is safe to irrigate with greywater. If you see water in the hole, the groundwater table is too shallow for greywater irrigation in that area.

In places with seasonally high groundwater, use greywater only during the dry, irrigation season, and divert it back to the sewer in the rainy season. (Homes on septic systems may also be legally required to divert back during the rainy season, though this practice is not protective of groundwater because septic leach fields discharge deeper underground than do greywater irrigation systems.) To protect drinking water wells, don't irrigate too close: codes often require 50 to 100 feet of horizontal distance between greywater irrigation and the well.

CHAPTER 5

Plants and Irrigation

Plants are at the heart of greywater irrigation systems.

To get the best results and save the most water, you'll need to choose appropriate plants to irrigate as well as calculate how much water they need. This chapter will teach these important details so you can best design the irrigation portion of your system. I'll teach you a few methods to calculate plant water requirements so you can adequately irrigate your plants and maximize your water savings. Last, I'll discuss what soaps and products are best to use in a greywater irrigation system to ensure the water is a good quality for your plants.

IN THIS CHAPTER:

Choosing Plants for Greywater Irrigation

A well-designed system finds a balance between the amount of greywater available and the irrigation needs of the plants. Since the amount of greywater and plant water needs both fluctuate, your design goal is to find an optimal match: irrigate as many plants as possible while keeping them healthy. During rainy times when your plants don't need irrigation, either turn off the greywater system or, in well-draining soils, keep it on. The amount of greywater going through a system is minuscule compared to a rainstorm.

If you have an existing landscape, follow these steps:

1. Decide what area of the landscape is easiest to direct greywater to. Be open to changing the landscaping, if necessary.
2. Determine if the plants in this area are appropriate to irrigate with greywater.
3. Estimate how much water the plants require (see page 49). Remember, this is estimating their peak irrigation needs; they don't need this much water most of the year.
4. Compare the amount of greywater available with the irrigation needs of the plant. Do they match up? Try to stay within 30 percent of plant needs. You may find your plants thrive with less water, and if they occasionally become water stressed you can supplement with rain or tap water.

Greywater can irrigate a variety of plant types.

MAXIMIZE YOUR WATER SAVINGS

If water savings is your goal, you must replace potable water irrigation in your landscape with greywater irrigation. If you have an existing irrigation system, try to replace an entire zone with greywater and then shut off the zone. If you can't replace a zone, be sure to shut off any emitters or spray heads that reach the greywater-irrigated area.

Installing a greywater system is a great time to assess your entire landscape. If you're designing a new landscape or redesigning an existing one, be sure to design the garden so that plants requiring frequent irrigation are near a greywater or rainwater source and, once established, the remaining plants thrive without irrigation. Implementing other water-wise landscaping techniques, such as choosing plants adapted to your climate (without supplemental irrigation), using mulch to prevent evaporation, and grouping plants with similar water needs (called **hydrozoning**), facilitates water-efficient irrigation. If you'll be irrigating some of your landscape with potable water, be sure to use the most efficient type of irrigation system you can; there are both low- and high-tech options (See Resources).

Greywater-Compatible Plants

Larger plants are better suited for greywater irrigation than smaller ones. A tree or bush with a large root area can withstand fluctuation in water much better than small plants can. Large plants also need more water than small ones, making it easier to distribute more greywater to fewer plants. As you look at your landscape, identify the easiest plants to irrigate. Most houses exhaust their greywater supply before the entire landscape is irrigated. If you end up with extra greywater, consider planting something new.

Some landscape areas aren't well suited for greywater irrigation, such as lawns or areas full of small plants (although high-tech systems can irrigate these types of plants; see page 173 for more info). Consider these techniques to improve a landscape for simple greywater irrigation:

- Remove a section of the lawn and plant perennials. Or, remove a strip of lawn around the edge for new plantings.
- Plant more greywater-compatible plants in the area. Consider taking the other plants off irrigation and, if they die, replace them with something suitable to your climate.
- Redesign the landscape so it is compatible with greywater irrigation (see Resources for ecological landscaping ideas).

Note: If you're wondering whether hot water from a shower or washer may harm the plants, don't worry; it won't. By the time hot water flows down the pipes, soaks through the mulch, and reaches the roots of plants it's not hot anymore.

These are the easiest plants to irrigate with greywater:

TREES. Fruit trees (or any trees) adapted to your local climate thrive with greywater irrigation.

BUSHES AND SHRUBS. Bushes and shrubs suited to your region are easy to irrigate with greywater. Consider fruiting varieties, or find ones that create bird and beneficial insect habitat.

VINES. Edible vines, like passion fruit or kiwi, are attractive and produce fruit.

LARGER PERENNIALS. Perennial vegetables, which produce year after year without needing replanting, are a productive addition to any landscape (see Resources for ideas). Flowering plants provide bird and butterfly habitat.

LARGE ANNUALS. Large annual plants, both edible and non-edible, can be irrigated with an L2L or pumped system; for example, tomatoes, corn, zinnias, squash. (Remember, you can safely irrigate food crops so long as the edible portion is above the ground and greywater doesn't touch it.)

SMALLER PLANTS GROWING CLOSELY TOGETHER can be irrigated in the middle of the planting area, so their roots share the water. Or, create distribution channels to move greywater toward the plants, like a "sun" mulch basin.

This apple tree thrives with greywater irrigation.

Passionflower vines produce beautiful flowers, provide delicious fruit, and grow over a large area.

"Sun" mulch basin. Greywater flows out from center of basin toward smaller plants.

TREES REDUCE HEATING AND COOLING NEEDS OF YOUR HOME

Use greywater to reduce your home's energy needs by growing trees. Deciduous trees, planted on the south and west sides of your home, block the summer sun to keep your home cooler, while letting in winter sun to warm your home. Trees do more than just reduce energy use by blocking the sun: as they evaporate water, heat is removed from the air — a natural air conditioner.

Trees can shade driveways and patios, preventing the hardscape from absorbing the sun's energy (which radiates back in the evening), keeping your outside environment cooler in the summer. Evergreen trees can block wind and keep your home warmer in the winter. Trees also block noise and glare, reduce pollution, and create an ecosystem for birds and other creatures. Before planting a new tree be sure to observe the sun patterns on your site so you can strategically locate the tree for the maximum benefit. See Resources for more information.

PLANTS FOR ECOLOGICAL DISPOSAL (WETLAND PLANTS). If you have ample irrigation water and don't need to be water-conscious in your landscape, consider growing water-loving wetland plants; they thrive with frequent and plentiful greywater irrigation. Or, if a lush wetland is in your garden design, dedicate some of your greywater for it. Note: It's a lot easier to direct a portion of the greywater to irrigate a wetland than to flow all the greywater through it before an irrigation system. (Wetlands are used to process greywater in places without irrigation need or sewer/septic options, and these designs flow all greywater through the wetland.) Backyard wetlands are prone to clogging, which prevents greywater from passing through.

NATIVE AND LOW-WATER-USE PLANTS. Use greywater to irrigate drought-tolerant and native plants, but be careful not to over-irrigate them. These plants can survive typical droughts in their climate, but they may look better during the dry times with a little extra water — the reason many people irrigate them. Design a greywater system to spread out water as much as possible through the landscape. Note that some native plants don't do well with summer irrigation (oak trees, for example). If you're looking for ideas for appropriate native or low-water-use plants for your climate, visit the EPA's website to find local sources of information (see Resources). Water districts and local extension services may also provide this information.

TELL YOUR LANDSCAPER OR GARDENER ABOUT YOUR GREYWATER SYSTEM

Anyone working in a landscape with a greywater system should understand how the system works and where the components are located; otherwise, they may unintentionally damage the system. Show your landscaper or gardener photos of pipe and outlet locations (pre-burial) and make sure they understand the importance of maintaining the outlets and mulch basins, never covering them with soil.

How Much Water Do My Plants Want?

Many people have no idea how much water a plant needs. Here is a rough estimate of how many fruit trees you could irrigate from a simple washing machine system, assuming there is no rain to supplement the greywater irrigation. Determine your irrigation potential by multiplying the first number (one load per week) by the number of loads you do each week.

COOL CLIMATES (65–75-degree summers)

- Front-loading machine (one load a week): 1 to 2 trees
- Top-loading machine (one load a week): 3 to 4 trees

WARM CLIMATES (75–85-degree summers)

- Front-loading machine (one load a week): 1 tree
- Top-loading machine (one load a week): 2 to 3 trees

HOT CLIMATES (85–100+-degree summers)

- Front-loading machine (one load a week): ½ tree
- Top-loading machine (one load a week): 1 to 2 trees

With the above estimates in mind, perform some basic calculations to determine more accurately how much to irrigate your plants. There are many factors that affect plants' water requirements, including climate, exposure (e.g., southern or northern), wind, shade, mulch, type of plant, and plant size.

Next, I'll show you two methods for estimating how many gallons per week a specific plant should receive. The first is a quick, rule-of-thumb estimate and is accurate enough to keep your plants healthy and happy. Use this method to guide your system design. For those of you wanting more detail, the second method (which incorporates evapotranspiration, or

ET, rates) can help you fine-tune your design, though this level of precision is not essential and the calculation is considered optional.

Keep in mind that these methods, as with any technique for determining irrigation needs, provide only *estimated* amounts. Always observe your plants: they can receive a wide range of irrigation amounts and still grow healthily, so long as the soil doesn't become waterlogged. Try to maximize your greywater potential and irrigate as many plants as possible; you'll save more water if you under-irrigate and add supplemental water occasionally, as opposed to consistently over-irrigating.

Find the area under a tree using $A = \pi r^2$. Area of tree's footprint = 3 × 4 ft. × 4 ft. = 48 square feet.

Rule-of-Thumb Estimate of Weekly Irrigation Need

Determine the weekly irrigation need of a plant based on its size and the climate. The plant size is measured by the area under its canopy; for trees and bushes, this area is shaped like a circle. Use the area (A) of a circle: multiply π (3.14, or round to 3.0) by the circle's radius (r) squared ($A = \pi r^2$). Planted beds or hedgerows have a rectangular area. Find the area of a rectangle by multiplying the length by the width.

TO DETERMINE THE PEAK IRRIGATION NEED (in gallons per week), first find the number of square feet of planted area, then divide by:

- 1 in a hot, arid climate
- 2 in a mild climate with warm summers
- 4 in a cool climate with coastal summer fog

For example, here's the calculation for an apple tree that measures 4 feet from the center of the trunk to the outer branches:

Area of circle = $\pi \times r^2$ (rounding π to 3) = $3 \times 4^2 = 48$ square feet.

Next, divide 48 square feet by the number for the climate:

- in a hot, arid climate: 48 ÷ 1 = 48 gallons/week
- in a mild climate with warm summers: 48 ÷ 2 = 24 gallons/week
- in a cool climate with coastal summer fog: 48 ÷ 4 = 12 gallons/week

This rule-of-thumb estimate is most accurate for plants with moderate water use, such as fruit trees. Water-loving wetland plants would like more water, while drought-tolerant or low-water-use plants require less. If you are irrigating a low-water-use or drought-tolerant plant, divide your estimate (gallons/week) in half again.

Estimating Water Requirements Using Evapotranspiration Rates (ET)

Evapotranspiration, or ET, is the combined effect of water evaporating from the soil and being used (transpired) by plants. In relatively warm, dry climates plants lose more water than in cooler, moister climates. This method assumes that all moisture by evapotranspiration will be replaced though irrigation. It factors in the type and size of the plant as well as the climate. You'll need the following information:

1. **THE AREA OF THE PLANT(S)** in square feet (see page 50).
2. **THE SPECIES FACTOR OF THE PLANT(S).** The plant species factor is high (0.8), moderate (0.5), or low (0.2) water use. Garden books typically provide this information and often represent it with a water droplet; a full drop is high, half-full drop is medium, and empty drop is low. Most fruit trees are moderate-water-use plants. If you can't find this info in a general reference book, look in The Water Use Classification of Landscape Species (WUCOLS; see resources) or on the website waterwonk.us.
3. **THE ET RATE FOR YOUR AREA.** The rate depends on climate conditions. Typically, ET is given in inches per month or per day. This can be converted to inches per week, which is what's needed for the formula. You can find ET rates online or ask your local cooperative extension program or water district.

The EPA's WaterSense website has a tool you can use to find your peak irrigation month and subsequent reference ET rate (see Resources). The tool also provides average rainfall for the peak irrigation month so you can adjust your irrigation estimate accordingly. Visit their website and simply enter your zip code. Keep in mind that this data is for the peak irrigation month and you don't need to irrigate this much all year long. Usually if you stay within 30 percent of this number, either lower or higher depending on your available greywater, the plants should be happy without supplemental irrigation.

Use the formula below to calculate weekly plant water requirements:

WEEKLY PLANT WATER REQUIREMENT

0.62 *(conversion coefficient to change inches/week to gallons)* × **ET** *(weekly)* × **species factor** × **plant size** *(square foot)* = **gallons per week**

The following examples look at irrigation requirements for a living fence made of pineapple guava bushes (moderate-water-needs plant; species factor: 0.5) covering a 100-square-foot area, using peak irrigation (as reported by the EPA WaterSense website).

These estimates don't take into account summer rains; obviously plants won't require as much water during the week if it rains. You can subtract average rainfall from the ET rate for this calculation. Depending on the rainfall frequency, however, this rain could come in one heavy storm or be spread out evenly over the month, and this will impact how much water plants would like during the week.

LOS ANGELES, CA

Peak ET, July: 6.73 inches/month or 1.68 inches/week:

0.62 × **1.68** *(inches per week)* × **0.5** × **100 sq ft** = **52 gallons per week** *(irrigating within 30 percent of this would be 36 gallons/week)*

TUCSON, AZ

Peak ET, June: 12.42 inches/month, or 3.1 inches/week:

0.62 × **3.1** *(inches per week)* × **0.5** × **100 sq ft** = **96 gallons per week** *(irrigating within 30 percent of this would be 67 gallons/week)*

SEATTLE, WA

Peak ET, July: 5.42 inches/month or 1.36 inches/week:

0.62 × **1.36** *(inches per week)* × **0.5** × **100 sq ft** = **42 gallons per week** *(irrigating within 30 percent of this would be 30 gallons/week)*

MIAMI, FL

Peak ET, April: 6.65 inches/month or 1.66 inches/week:

0.62 × **1.66** *(inches per week)* × **0.5** × **100 sq ft** = **51 gallons per week** *(irrigating within 30 percent of this would be 36 gallons/week)*

A GREYWATER POND?

My first two greywater systems flowed through a bathtub wetland into a greywater pond, overflowing with water hyacinth. I loved the 8-foot-tall cattails that filtered shower water clear and odorless. It took a few years before I realized this wasn't a greywater reuse system, rather an ecological disposal system; the cattails were sucking up the water that could have been irrigating my thirsty garden. (Once mature, the plants literally sucked up an entire shower on a hot day — not a drop overflowed out of the wetland to the garden.)

Though technically easy and also beautiful and fun, treating greywater for a backyard pond is not something I recommend. If your goal is to save water, a greywater pond isn't the right choice. The quality of greywater is unsuitable to fill a pond; it must be filtered by wetland plants first (unless you want a pool of disgusting, stinky greywater). Thirsty wetland plants not only use water, but they also remove nutrients and organic matter from the water, which your garden could benefit from; less greywater flows out of the wetland than in.

A water-wise choice for anyone with both a pond and a landscape is to irrigate with greywater and fill the pond with freshwater (or rainwater), bypassing the need for the water-loving wetland filter. Separately, greywater ponds epitomize the concerns of the regulatory world. Their list of potential hazards includes: drowning risk for children, potential for direct contact with the water, mosquito breeding grounds, and possible overflow to a neighbor's yard or storm drain. Ponding greywater is prohibited by even the most lenient codes.

Permaculture centers take pride in their greywater ponds. Shining photos of rural greywater-fed ponds are common in blogs, books, and magazines. It's not often emphasized, however, that these ponds are typically filled with rainwater while just a fraction is "grey."

People who try to replicate them are disappointed when their backyard greywater pond is a slimy, stinky, algae pool. A backyard rainwater pond won't cause such headaches! (See Resources for info on rainwater ponds.)

Plant-Friendly Soaps

Greywater can either benefit or harm plants, depending on what soaps and detergents you use. Its quality as an irrigation source is directly connected to what you put down the drain. Luckily, it's easy to choose soaps and other products that are plant-friendly, avoiding the following ingredients:

- **SALT AND SODIUM COMPOUNDS.** Salts can build up in the soil and inhibit plants' ability to take up nutrients and water. Minimize and avoid salts.
- **BORON.** This plant microtoxin is damaging even in small amounts. Do not use any products that contain boron, including the laundry additive borax. Because it is nontoxic to people, boron is found in many ecological products.
- **CHLORINE BLEACH.** Bleaches containing chlorine kill microorganisms, including beneficial soil microbes. Hydrogen peroxide bleach can be used instead, or you can turn off your greywater system when using bleach.
- **ALKALINE COMPOUNDS (OPTIONAL).** Some products raise the level of pH, making the water more basic (or alkaline). This isn't a problem for most plants, although some types (such as blueberries and azaleas) prefer acidic conditions, and basic water may not suit them. In general, liquid soaps do not increase the pH of the water, whereas bar soaps do. Cleaning

products can also be extremely basic (alkaline). If you are using greywater from a source where only liquid, pH-neutral products are used, greywater can irrigate any plants, including acid-loving varieties. Refer to garden books, extension offices, or local nurseries to determine whether your plants are acid-loving.

Product Recommendations

Following are some products that have been used successfully for many years in greywater systems. This list is not exhaustive, and you may find others that are free of boron and very low in salts. Additionally, you can look up the ingredients for personal care products on the Environmental Working Group's *Skin Deep Cosmetics Database* (see Resources).

- **Washing machine:** ECOS, Bio Pac, Oasis, Vaska, Puretergent, FIT Organic, as well as non-detergent options like soap nuts or laundry balls. Powdered detergents are never okay; use only liquid detergents. Watch out for brands like 7th Generation that claim to be greywater-safe but contain boron and salts.
- **Showers:** Aubrey Organics (most types), Everyday Shea, Dr. Bronner's. In general, typical shampoos and conditioners will not harm your plants. The products are very diluted, liquid (very low in salt), and free of boron.
- **Sinks:** Oasis All-Purpose Cleaner, Dr. Bronner's Pure-Castile Liquid Soap, most glycerin-based soaps. Cleaning products: Use vinegar-based products, not white powders. Or, turn off the greywater system if you need to do a deep scrub with a salt-based "powder" cleaner.

SALTS

The amount of salts you can send into your yard without damaging your plants depends on your climate, soil, and plants. If you live in a place with heavy, frequent rainfall, rain will leach salt out of the soil before it can build up to harm plants, so the occasional salty product won't cause any harm. On the other hand, in places with salty tap water (such as groundwater or Colorado River water) and a dry climate, soils are more prone to salt buildup, so you should take more care to avoid adding salts from greywater. And keep in mind that fertilizers are high in salts and that salt tolerances of plants vary considerably. In arid climates, direct rainwater into greywater basins as well as rainwater basins to flush salts from the soil.

CHAPTER 6

Choosing a Greywater System

The next step is to review your site and overall goals to determine which greywater systems are the best match.

In this chapter we'll review basic design considerations, where to send the water, and when not to use greywater. I'll discuss how to use greywater indoors for toilet flushing, and, for renters or others who don't want to alter their home, how to use greywater without a system. This chapter ends with a review of greywater systems types and their general costs, capabilities, and skill level required to install them.

IN THIS CHAPTER:

System Design Considerations

What part of the landscape should you irrigate — the lawn uphill from your house or the fruit trees downslope? Before getting to specifics, I'll cover general design principles to help narrow your options and define your goals. Start with the following basic considerations.

CAN YOU HAVE A GRAVITY IRRIGATION SYSTEM? First, consider areas of your yard close to the greywater source that require irrigation and are NOT uphill or across large areas of hardscape. Gravity-based systems typically last longer, cost less, use fewer resources, and require less maintenance than pumped systems.

DOES WATER INFILTRATE WELL IN THE AREA YOU PLAN TO USE GREYWATER? If this is an area you don't already irrigate, you should perform an infiltration test (page 41) to make sure greywater will soak into the soil without pooling.

WILL THE LOCATION OF THE GREYWATER SYSTEM CAUSE ANY UNFORESEEN PROBLEMS? Maintain setbacks to the house's foundation, neighbors' yards, retaining walls, streams and lakes, groundwater table, and water supply wells. See the chart at right for some standard setback distances (based on California code). Setback requirements vary by area; check with your local code authority for a permitted system. Codes often include setbacks for septic tanks and leach lines, although it seems illogical to require a setback from a leach line since greywater would have been in the leach line if it wasn't diverted from the septic system.

EXAMPLE SETBACKS FOR GREYWATER IRRIGATION AREA *(minimum distance)*

Building foundations	2 feet
Property lines	1.5 feet
Water supply wells	100 feet
Streams and lakes	100 feet
Water table	3 feet above
Retaining wall	2 feet

In addition to the basic elements shown in the chart, greywater surge tanks also may have setback requirements. If your project requires a permit and you need a tank, ask your local building department for their thoughts on tank placement. Other cases may involve special situations that require modifications to the setback requirements. For example, if your neighbor's yard is lower and covered with hardscape (e.g., a concrete patio), a 1.5-foot setback will not be enough distance to prevent leaking of greywater onto the neighbor's patio. Increase setbacks anytime the greywater-irrigated landscape is elevated above any type of hardscape, like sidewalks, patios, walkways, or roads.

AVOID COMMON PROBLEMS AND DESIGN PITFALLS

→ **DON'T STORE GREYWATER!** Nutrients in greywater break down over time, causing a stink. To avoid odor problems with surge tanks (in a pumped system), make sure the tank is vented, can be cleaned, and is located out of the living space. (And don't send greywater to mix with rainwater in a tank; the tank will become contaminated.)

→ **DON'T ALLOW GREYWATER TO PUDDLE OR POND UP.** It is non-potable and people shouldn't contact it. Ponding also creates mosquito breeding grounds.

→ **KISS (KEEP IT SIMPLE OR YOU'LL BE SORRY).** The more components and parts used, the more potential for system failure. At the residential level, simple systems tend to work best, last longest, and cost less than complex setups. Many systems do not require a tank, pump, or filter; use these components only when necessary.

→ **PLAN CAREFULLY** to match your greywater production with the needs of your plants.

→ **USE PLANT-FRIENDLY PRODUCTS** (those without lots of salts, boron, or chlorine bleach).

DESIGNING A NEW HOUSE WITH A GREYWATER SYSTEM

If you plan your house with the greywater system in mind (not to mention rainwater catchment and waterless toilets), you'll have more options and an easier system to install. Here are some planning tips:

- → Locate the house above the landscape area when possible.
- → Keep the plumbing accessible (no concrete slab). If you will have a slab foundation, design an accessible location for diverting greywater, and don't join greywater and blackwater pipes below the slab.
- → Design the landscape to fit with your greywater (and rainwater) supply.
- → If you are not planning to install the greywater system immediately, put in the diverter valve with a "greywater stub-out," or create a place to easily add it in later.

See page 22 for more details on greywater access locations.

When Greywater Is Not a Great Idea

Outdoor greywater irrigation systems are not always appropriate. Here are some situations where greywater reuse may not be suitable:

UNINTERESTED AND UNCOOPERATIVE PEOPLE LIVING IN THE HOME. If people won't use plant-friendly products or maintain the greywater system, it could harm the landscape (unless the landscape is designed with salt- and boron-tolerant plants, and an outside person does the maintenance).

DIFFICULT-TO-ACCESS DRAINPIPES. When it's very difficult to access greywater sources, such as a house on a slab foundation with all greywater sources in the middle of the home, money may be better spent on an alternative system, such as ultra-efficient fixtures and appliances and rainwater harvesting systems.

VERY POORLY DRAINING SOIL. If you live in a swamp, it will be hard to infiltrate greywater into the native soil, and you won't be saving water; swamps don't need irrigation. Consider an ecological disposal system, like a constructed wetland, or an indoor greenhouse greywater system (see page 167–169).

If you can't use greywater for irrigation, investigate options for indoor reuse, such as for toilet flushing, or focus on rainwater catchment or composting toilet systems.

NOT ENOUGH SPACE FOR PLANTS AND IRRIGATION SYSTEMS. With a very small (or nonexistent) yard, there may not be enough landscape area to infiltrate greywater.

TOO CLOSE TO A CREEK OR A DRINKING WATER WELL. Setback requirements range from 50 to 100 feet.

UNSTABLE SLOPE. Adding extra water to a steep or otherwise unstable slope could cause a landslide.

Using Greywater Indoors: Toilet Flushing

Flushing the toilet with pure drinking water offends common sense and ecological awareness. Why not flush with non-potable greywater instead? In the right situation, flushing with greywater is a great idea. However, in many situations there are better ideas to implement first. It's typically cheaper and easier to set up an outdoor greywater irrigation system. It's also easier to install a rainwater system to flush the toilet than a greywater one. But if your site doesn't need irrigation, there are systems that filter greywater to flush the toilet.

An integrated toilet tank lid sink allows you to wash your hands with the clean water as it fills the toilet tank. There are options for a retrofit sink lid (called "SinkPositive"), compatible with existing toilets, or a new toilet with the sink built in, like this Caroma Profile 5 Toilet Suite Deluxe with an integrated hand basin.

Challenges with Toilet Systems

Toilet-flushing greywater systems usually require frequent maintenance, manual cleaning of filters, and chemical disinfectant to prevent odors in the bathroom. They also tend to be relatively complicated, and it's critical that they be designed and installed properly. A few companies sell systems, while some greywater professionals design their own. Studies and system users report that lower-cost systems ($3,000–$5,000) have maintenance issues, while the better-functioning versions have a steep price tag ($8,000–$9,000). In general, it costs far less to purchase an ultra-low-flow toilet and showerhead than invest in a toilet-flushing system.

Getting a permit tends to be more difficult for indoor reuse of greywater, although a few states, such as Florida, make it easier than outdoor reuse. Some tinkerer-types create functional mechanisms for flushing, but these are never up to code. A study on 25 toilet-flushing systems in Guelph, Canada, found an average savings of just 4.5 gallons per person per day (gpcd). Toilets using treated greywater tend to require more maintenance, and internal parts wear out, causing wasteful leaks.

Recommendations

Use shower water first for indoor systems, since it has fewer particles to filter out. You'll need to access the drainpipes to install the diverter valve, and install a tank with a filter. Lastly, do a reality check to make sure you are ready to do the necessary maintenance for this type of system to function well.

Choosing a Greywater Irrigation System

Greywater is a unique type of irrigation water, distinctly different from rainwater or potable water from the tap. To choose the best greywater system for your situation, you'll need a basic understanding of how different types of systems work and their advantages and limitations, as well as a clear understanding of your site: the landscape, plumbing configuration, your budget, and how much maintenance will realistically be done to upkeep the system. Much of this information you gathered in chapters 1–5, and the remaining considerations are covered here.

Combine Greywater Sources or Keep Them Separate?

One big consideration is whether to keep the greywater flows separate and install a system for each fixture or to combine the flows together and install one larger system. For example, will you combine the showers with the washing machine greywater or keep them separate? What's best depends on your situation: the layout of your home, the type of system you want to install, and what types of plants you plan to irrigate. In general, if you are installing a system in your current home and are not doing a major plumbing remodel, it's most practical and economical to keep the greywater sources separate and build a different system from each fixture. In new home construction, or during big remodels, you could combine flows together and install one larger system.

WHEN SEPARATE SYSTEMS MAKE SENSE

- Your house is already built and you don't want to do extra plumbing work.
- You want a simpler and lower-cost project.
- You plan to use gravity systems for showers/sinks and a laundry-to-landscape system for the washing machine.
- You need irrigation water near the various greywater sources. For example, one shower is close to the front yard, while the washing machine is near the backyard.
- **Tip:** Even if you combine some flows, consider keeping the washing machine separate. Having a separate valve next to the machine allows you to control the laundry flow without having to shut off the entire system.

WHEN COMBINING FLOWS MAKES SENSE

- You plan to incorporate a pump or filter.
- You want to include fixtures that aren't used frequently, such as those in a guest bathroom.
- You want multiple irrigation zones. You'll typically need more than one fixture to generate enough water for multiple irrigation zones.
- **Tip:** If you plan to reuse your entire household's greywater and have multiple bathrooms in the home, consider keeping one bathroom off the system. Guests can use that bathroom without sending their own (possibly non-plant-friendly) soaps to your yard, or forcing you to turn off the entire system.

Other Considerations: Permits and Your Existing House

Permits and their requirements are other significant factors. Some types of systems are much easier to get a permit for than others, and many local authorities impose rigid permitting guidelines, which will impact your system design. In addition, most homes and landscapes were not planned with a greywater system in mind, so the best greywater system for your site may not be quite what you originally imagined. You may discover your site presents logistical challenges or your landscape plants aren't suitable. Being flexible and open to change, particularly landscape changes, will help you find an affordable and suitable system.

Greywater Systems at a Glance

In the upcoming chapters I'll explain how to design and install four common systems: the landscape-to-laundry (L2L), branched drain, and two pumped systems (both with and without a filter). The Greywater Systems Overview chart on page 63 summarizes the costs and capabilities of these systems.

LAUNDRY-TO-LANDSCAPE SYSTEMS typically are the lowest in cost and easiest to install. Because the installation doesn't alter the household drainage plumbing the system often doesn't require a permit.

BRANCHED DRAIN SYSTEMS are gravity-based and require the landscape to be lower than the plumbing; these are best suited for larger plants. They're very low-maintenance — a great choice for irrigation of trees and bushes.

PUMPED SYSTEMS, both filtered and unfiltered, are designed to move greywater uphill and are capable of spreading the water over large areas. These systems typically are more difficult to get a permit for and may require backflow prevention (see page 150). Check with your local building authority, as specific requirements can add significant costs to the system.

Which System to Do First?

A common order of installing systems — for people retrofitting existing homes, as opposed to new construction or major remodels — is to start with the laundry-to-landscape system, then install gravity or pumped systems from the other fixtures. A kitchen sink system requires more frequent maintenance than other sources of greywater due to higher organic matter in the water, so it's best to include one after the easier-to-maintain systems are installed.

GREYWATER SYSTEMS OVERVIEW

Type of system	Laundry-to-landscape (L2L)	Branched drain	Pumped — no filter	Pumped — with filter
FIXTURES	Washing machine only	Showers/baths, sinks, kitchen sink, washing machine	All fixtures	All fixtures (kitchen sink typically is not recommended
MATERIALS COST *(excluding permit)*	$100–$300	$250–$500	$500–$1,500	$1,000–$5,000
COST OF PROFESSIONAL INSTALLATION *(excluding design fees for any permits)*	$700–$2,000	$800–$4,000	$1,500–$4,000	$2,000–$15,000
SKILLS NEEDED FOR TYPICAL INSTALLATION	Basic landscaping and construction; basic plumbing	Basic landscaping, construction, and plumbing. Plumbing can be complex.	Basic landscaping, construction, plumbing, and sometimes electrical. Plumbing can be complex.	Basic landscaping, construction, plumbing, and sometimes electrical. Plumbing can be complex.
ABLE TO MOVE WATER UPHILL?	No more than a few feet	No	Yes	Yes
SUITABLE FOR TREES AND SHRUBS?	Yes	Yes	Yes	Yes
SUITABLE FOR SMALLER PLANTS?	Yes, though front-loading machines can't distribute water to many locations.	Not usually. Possibly if plants are clustered and irrigated in the center of the group.	Yes, though not able to spread water out as far as with drip irrigation.	Yes; also suited to drip irrigation.
SUITABLE FOR LAWN?	No	No	No	Possible, though more challenging.
MAINTENANCE	Annual	Annual	Annual	Every 1–2 months, possibly less often (the frequency depends on the site and specifics of the system). Self-cleaning filtered systems: annually

OPERATION AND MAINTENANCE MANUAL FOR YOUR GREYWATER SYSTEMS

You will need an Operation and Maintenance (O&M) manual for each greywater system (and most codes require you to have one). This manual includes important information about your greywater system, including:

- What types of soaps and detergents can be used
- When the system should be shut off
- How much greywater it was designed for
- Maintenance requirements
- Basic troubleshooting tips
- A site plan of the system accompanied by photographs of the unburied pipes; this will help you find pipes in the future in case you want to alter the system or plan to do landscape work and don't want to damage the pipes.

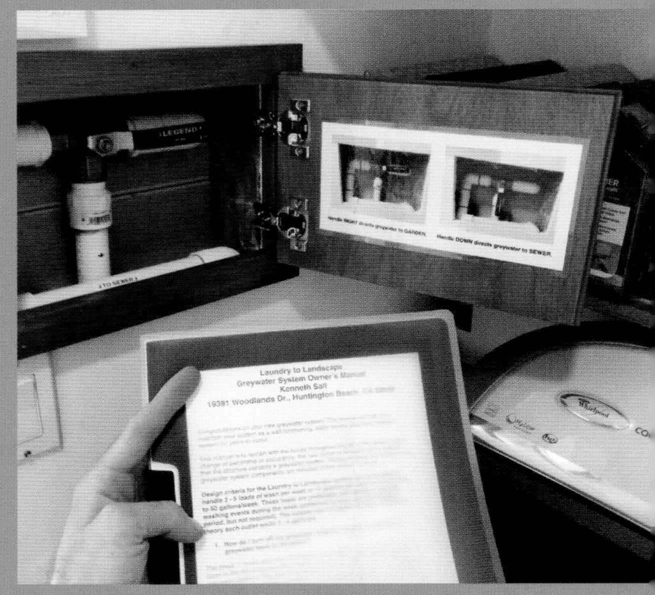

Place the O&M manual in an easy-to-find location, such as inside a plastic sleeve and taped near the diverter valve or greywater fixture (for example, on the side of the washing machine). Show the manual to any house-sitters or house guests, and pass it on to new owners if you sell the house. Some local municipalities, either the city or water district, can provide you with an O&M manual to adapt; or you can download one from the website of a nonprofit such as Greywater Action (see Resources).

When to Turn Off the System

If you have a functional sewer or septic system, it's easy to turn off your greywater and direct the water to the sewer/septic system. Direct the greywater to the sewer/septic system when:

- Using chlorine bleach, hair dye, harsh cleaners, or other toxic substances
- Using non-plant-friendly products (those high in salts or containing boron)
- Washing to remove toxic or hazardous chemicals from clothing, such as gasoline or paint thinner spilled on your shirt
- Washing diapers or clothing containing fecal matter
- There are puddles in the yard or any visible signs of greywater surfacing (note that some jurisdictions require systems to be shut off at the start of the rainy season)

CALL BEFORE YOU DIG!

Call 811 (or visit www.call811.com) several days before starting a project to find out where your underground utilities are located. The hotline will route your call to your local utility call center. You'll tell the operator where you're planning to dig and they'll notify the utility providers in your area, who in turn will mark any underground utility lines on your property. You can also hire a private locator to find lines inside the property; this might be advisable if you have outbuildings, such as sheds and guest houses, that may be supplied by electrical and water lines from your house but aren't officially within the purview of utility companies.

In addition to having all utilities marked, it's up to you to make sure you don't put a pickax through a buried service line, which could be a deadly mistake. Use extreme caution when working anywhere near buried lines. Or better yet, avoid these areas entirely.

USING GREYWATER WITHOUT A SYSTEM: TIPS FOR RENTERS AND HOMEOWNERS

It would be great if all landlords were water-conscious and supported the efforts of their water-wise tenants. Some are, but many still don't want tenants changing the house, which makes installing a greywater system more challenging. If you're a homeowner, perhaps you aren't ready to install a dedicated system or you just want some simple solutions for reusing greywater right away. Following are a few ideas for reusing water without altering the house. Don't forget to check with your landlord about these projects; they may want to pay for the parts or learn more.

"LAUNDRY DRUM" SYSTEM. Simply stick the hose of the washing machine out a nearby window (nearby window required), drain it into a drum/small tank (about 50 gallons), then connect a garden hose to the bottom of the barrel and move the hose manually to water the plants (see page 67). Since greywater can't be stored you'll need to use it that day.

LAUNDRY-TO-LANDSCAPE SYSTEM WITH NO HOLE IN THE WALL. Attach a piece of wood, such as a 2 × 4, to the bottom of a nearby window and drill through the wood instead of the wall. The window closes against the wood and the pipe exits the house with no holes in the house. Alternatively, if your landlord allows a hole in the floor (and you have a crawl space), drill through the floor and exit the house via a screened crawl space vent.

OUTDOOR WASHING MACHINE. Set up your washing machine outside on a covered deck or porch (requires above-freezing temperatures). Connect the machine to an outdoor spigot, and run only cold-water loads (use a washer Y hose to connect cold water to both cold and hot inlets on the machine).

Laundry-to-landscape system set up on an outdoor washer.

BUCKET OR SIPHON. Collecting greywater in buckets builds muscle and provides a visceral awareness of how much water you use. Collect shower greywater by plugging the tub and to bailing out the water, or simply shower over a bucket. And don't forget to collect the "clear water" while your shower heats up. If the landscape is lower in elevation than your tub, you can siphon the water outside (with a hand pump and garden hose, or a product such as a SiphonAid). Kitchen sink water can be collected in a dishpan in the sink.

LAUNDRY DRUM SYSTEM

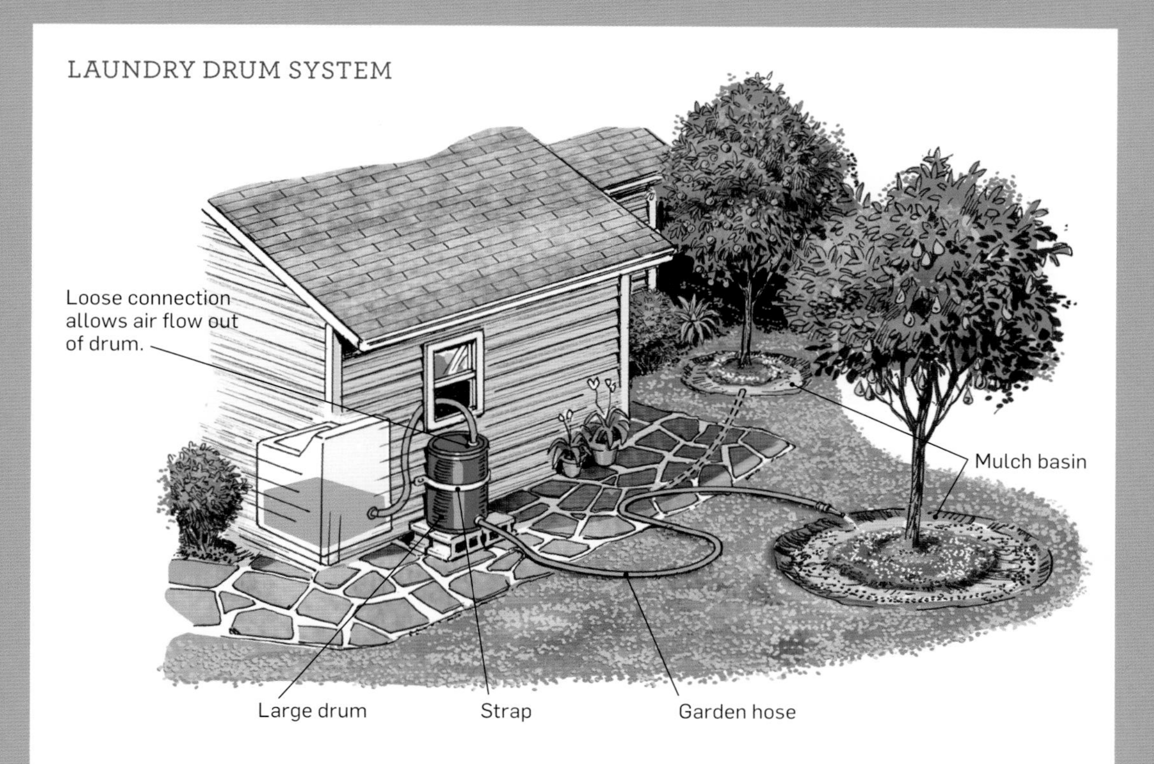

Constructing a laundry drum system is similar to converting a 55-gallon barrel into a rain barrel. The key difference is there is no shutoff on the bottom, since greywater should not be stored.

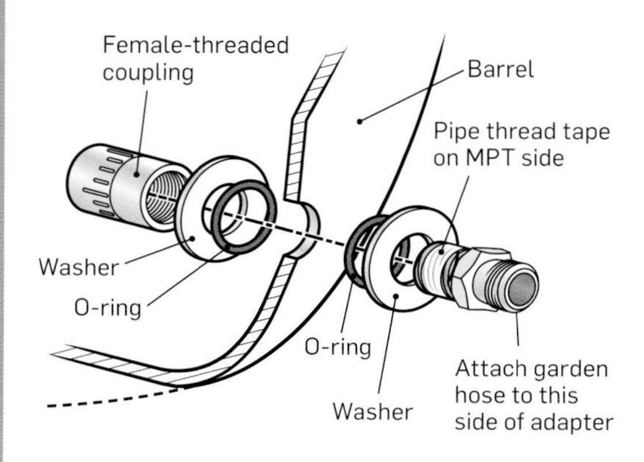

Directly attach a garden hose to the barrel and place the hose in a well-mulched part of the yard near thirsty plants to soak up the greywater. One method to attach the hose is to drill a hole at the bottom of the barrel, just large enough to thread a ¾-inch garden hose adapter or hose bibb into the barrel (use one without a shut-off to prevent accidental overflow of greywater). If you can't access the inside of the barrel, put a bead of silicone caulk around the hole to prevent leaks.

With access to the inside of the barrel you can more securely attach the fitting by doing the following; wrap pipe thread tape on the male pipe threads of the garden hose adapter or hose bibb, put an O-ring and washer onto the male threads and connect a ¾-inch female-threaded coupling to it, tightening with tongue-and-groove pliers. Put a bead of silicone caulk around the outside and inside to prevent leaks.

CHAPTER 7

Codes and Regulations

Understanding your local greywater code is important for a few reasons.

If your project requires a permit, understanding the perspective of your regulatory agency will help you work together. If you're working on policy change, you'll need to figure out how to have a functional code that simultaneously addresses concerns of health and safety officials. In this chapter you'll read about the diverse ways greywater is regulated around the country and learn some tips to improve your local codes from code-changing pioneers Greywater Action for a Sustainable Water Culture and Rob Kostlivy. You'll also hear perspectives from two water utilities that are promoting greywater.

IN THIS CHAPTER:

STATE OF CALIFORNIA PUBLIC HEALTH DEPARTMENT SUPPORTS GREYWATER REUSE

Public health officials supported California's revised greywater code. At the public hearing for adopting the 2009 greywater code (which eased permitting barriers for simple greywater systems), Dr. Linda Rudolph of California's Department of Public Health spoke in favor, "We believe that the proposed standards are adequately protective of public health and may provide substantial water conservation benefits and decrease demand on existing water supplies. We think this guidance will improve public health protection . . ."

A Brief History of Greywater Plumbing Codes

Historically, plumbing codes did not distinguish between greywater and blackwater (from toilets). Greywater was required to be collected together with blackwater and sent to the sewer or septic systems, and reusing greywater was illegal. This began to change in the early 1990s when drought-prone California realized this potential source of irrigation water was being wasted. The state plumbing code changed to allow legal reuse. Greywater advocates from that time will tell you how this code, though a positive first step, was practically useless. It treated greywater like septic water, requiring a small septic-type system to dispose of it deep underground (with a tank and gravel filled leach lines). People interested in irrigating with greywater still had to build illegal systems. California alone had an estimated 1.7 million unpermitted systems. States like Arizona, which followed California's example code, had a similar experience.

Arizona Breaks New Ground

In 1998 greywater pioneer Val Little, director of Water Conservation Alliance of Southern Arizona (Water CASA), conducted a survey in southern Arizona and found that 13 percent of residents used greywater, all illegally. The overly restrictive code prevented her from teaching people how to reuse greywater properly, so she worked to change the regulations. The end result was a performance-based code that outlines health and safety requirements (see Greywater Codes: Performance and Prescriptive on page 74). Residential greywater systems that follow the guidelines are legal — without permits, fees, or inspections — so long as the system produces fewer than 400 gallons per day. Now water departments, NGOs like Water CASA, and the state environmental health department can offer advice, brochures, classes, and financial incentives to encourage safe, legal greywater reuse.

Arizona's success was emulated by other states, including New Mexico and Wyoming. Eventually (in 2009), California upgraded its state code to remove some permitting barriers for irrigation systems.

The Current Situation

Greywater codes still don't exist in many parts of the country. At the time of writing, some states regulate greywater like septic water and require a septic disposal system for it. Others, like West Virginia and Massachusetts, allow greywater systems only in houses with a composting toilet. Florida bans outdoor greywater use but allows it for flushing toilets. Georgia allows you to carry greywater in buckets to the plants, but you can't get a permit to build a simple greywater irrigation system. Washington state's code allows very small systems built without a permit (following performance guidelines), but all other systems have quite stringent requirements. Oregon requires an annual permit fee.

Many barriers still exist for legal greywater systems around the country, but the tendency

The Environmental Health Perspective

Rob Kostlivy

Environmental health directors have a big job.

They regulate food production, solid waste, water supply, vector control, hazardous materials, milk and dairy products, air sanitation, noise control, rabies and animal disease — the list goes on. Greywater often falls under their charge, though most receive no formal training in it.

At the very first California Greywater Conference, Rob Kostlivy, Director of the Tuolumne County Environmental Health Department, gave the keynote welcome speech. He explained why environmental health departments have often blocked legal greywater reuse, and how he's working to change that. I spoke with Rob about his work in Tuolumne County. (Note: Environmental health departments have different names state by state, including variations such as Department of Ecology, and Department of the Environmental Quality.)

Why are environmental health departments often a barrier to legal greywater reuse?

The fear of the unknown has been a driving force that has plagued environmental health departments across the state for many years. We were taught that greywater is blackwater, virtually the same standard as septic system waste. Without the proper training and education, most jurisdictions chose to fear greywater in lieu of trying to understand it or become educated in the subject.

Tuolumne County has permitted a lot of greywater systems and sponsored the first California Greywater Conference. Can you talk about how your department is promoting safe reuse of greywater?

I looked at the literature from other places using greywater, like Australia. When I read reports from their environmental agencies, it helped me become comfortable with greywater. As a result, we permitted the largest greywater project in California at Evergreen Lodge, over 50 interconnected and independent systems at that facility. We worked closely with a local installation company to ensure health and safety precautions were in place, while simultaneously educating ourselves about different types of systems and reasonable safety precautions; we didn't want to force an over-engineered or overly costly system, but we still wanted to ensure that we protected public health. We've been monitoring the systems since 2009 to understand what is working and what needs to be improved. So far everything has been working well, and we've even permitted a kitchen sink system from the cafe.

To promote greywater, Tuolumne County partnered with Sierra Watershed Progressive to create California's first greywater conference. My intentions were simple: I wanted the *regulated* to sit across from the *regulators*. I knew we had a lot to learn from each other and felt it was important that the curriculum at the conference be conducted by the experts in the field, the grassroots organizers, educators; I also wanted the greywater community to hear the regulator perspective as well. I felt that we needed to work together to help develop resolutions in solving our statewide water crisis. I received feedback from my environmental health colleagues that this conference was a success and that they will continue to learn and work at promoting a smart, sensible and scientific approach to permitting greywater. Since then we've hosted a second conference and are planning a third. In my humble opinion, this is a great turnaround in attitude by the regulators. It shows how with partnership we can accomplish great things!

is toward better, friendlier codes. To find out how greywater is regulated in your state, refer to the state's plumbing codes or contact your local building department or environmental health department. Greywater is usually regulated by either the plumbing code (building department) or the environmental health department (or department of the environment). You can also contact the water district or environmental groups, though they may not be up to date on the code if a recent change has taken place.

HAVING TROUBLE GETTING A PERMIT?

Getting permits is important to mainstream greywater reuse, as well as to keeping your home up to code. Unfortunately, it's not always easy to get a permit, especially in places lacking precedent. Early adopters seeking permits work hard and often are forced to build a more costly and complicated system than necessary. Even in states lacking a greywater code, permits may be obtainable through the "alternative methods and materials" section of the code. Some homeowners have had success using their home as a "pilot" project that allowed local regulators to issue an experimental permit and then monitor and learn from that system.

Greywater Action

For a Sustainable Water Culture

Greywater Action (the group I helped start) educates about greywater, rainwater, and composting toilet systems through hands-on workshops and presentations.

Prior to 2009 our name was The Greywater Guerrillas, and all the greywater work we did was illegal, primarily due to an overly restrictive state code. After teaching hundreds of Californians how to install simple systems, we worked to legalize the practice. Responding to strong public support, millions of illegal systems, and a statewide drought, California changed its greywater law.

Workshop participants installing a branched drain system.

What advice do you have for other groups wanting to offer greywater education?
Start by gaining practical experience; install a system in your own home. People want to know what it's like to live with a system. How do the plants respond? Do greywater-friendly detergents really get the clothes clean? It takes time to gain experience, so the sooner you can start the better. In the meantime, collaborate with experienced people, attend trainings, and find partners to combine skills (plumbers, handy-people, permaculturists, landscapers).

How can people promote greywater in places without a code?
Regardless of legality, firsthand experience and working local examples are critical. Many regulators have never seen a greywater irrigation system, which makes it challenging to get a good code in place. Install systems based on designs proven in other regions as demonstration sites. Invite regulators for an "unofficial" visit to see them. Policy change requires effort at both the state and local level. Local regulators are hesitant to allow something not sanctioned by the state. Also,

Responding to strong public support, millions of illegal systems, and a statewide drought, California changed its greywater law.

you'll need buy-in at the local level to accept a new state code and promote it.

Here are some ideas based on what we've found successful:

- **BUILD A SUPPORT BASE.** In the two years leading up to our state's code change we led hundreds of people through the installation process of diverting laundry and shower water to irrigate the landscape, as part of our one-day workshop model. It was fun and empowering, and it connected the home to larger water politics and left an enthusiastic homeowner with a greywater system. These were the people who flooded state policy makers with letters, emails, and calls demanding that greywater be legalized. Regulators commented that the code change meetings were the largest they'd ever seen with the most public comments.
- **FIND AN INFLUENTIAL ALLY** to go through the permitting process (or pilot project if the codes haven't changed yet): a city council member, water department employee, or someone in the building department. Find the perfect site: very simple to install, nothing tricky or questionable, a model system. Ideally this would be at the home of your influential ally, or your own home.
- **LET THE CITY MONITOR THE SYSTEM** if it's a pilot project. Get lots of press. Conduct tours.
- **TO INFLUENCE STATE POLICY,** first find out how greywater is regulated in your state. If the department in charge isn't open to change, find a supportive politician. Though state legislation doesn't write codes, it can require the responsible agency to rewrite the code.
- **AFTER THE CODE CHANGES,** get permits for various types of systems and work with the building department to create a simple permit application and checklist for future permits. (And pressure the water district to incentivize greywater.)

Greywater Codes: Performance and Prescriptive

Performance-based codes describe health and safety requirements for greywater systems. Systems that meet the requirements are legal; those that don't are not. Performance-based codes often don't require inspections or fees for the lowest risk situations (for example, single-family, non-pressurized systems), yet provide legal grounds for a city to take action against a problem system. For example, "no pooling or runoff" is a common guideline that prevents exposure to greywater, but many codes don't specify how to meet this requirement. Performance-based codes are written in simple, straightforward language. States and local jurisdictions can provide further guidance, such as how to size a system to avoid pooling and runoff, but the more specific details are left out of the code.

Prescriptive-based codes specify exactly how to build a greywater system, including what materials and parts can be used. Instead of stating, "No pooling or runoff allowed," they may estimate greywater production based on the number of bedrooms in the house and size the irrigation area based on soil type.

A detailed code that spells out how to construct a greywater system will result in safer, better-built systems, right? Unfortunately, that's not the case. Greywater systems are complex; they interact with the living world of soils and plants, and are influenced by water-use habits, fixtures, climate, and the physical layout of the house and landscape. Unless the code considers all these variables (and, in fact, it never does), it results in overly restrictive requirements, adding unnecessary cost, or it creates an inefficient irrigation system. When a code is out of touch with reality, people ignore it and build illegal systems, with no guidance. After all, since it's common sense to reuse the water we already have, why should it be difficult to get a permit or the fees be expensive?

Label aboveground pipes containing greywater.

National Codes and Standards

Wouldn't it be great if there were just one code for the whole country, so each state didn't have reinvent the wheel around greywater law? The International Association of Plumbing and Mechanical Officials (IAPMO) writes codes that are adopted nationwide, and it has one on greywater. You can find it in the Uniform Plumbing Code (UPC) non-potable water chapter. Unfortunately, this code isn't very good. States that adopt it will need to alter it — as California did for its 2013 plumbing code — or risk minimal compliance from the public.

The International Code Council (ICC) also writes codes adopted by many states. Its greywater code, found in the International Plumbing Code (IPC) non-potable water chapter, is even worse than the UPC. (States that use the IPC should write their own code instead of adopting the greywater chapter.)

National standards for indoor reuse systems are being developed. NSF International recently released water quality guidelines as part of its standard for non-potable indoor reuse (toilet flushing). NSF 350: On-site Residential and Commercial Water Reuse Treatment Systems is, according to NSF, "a revolutionary standard that sets clear, rigid, yet realistic guidelines for water reuse treatment systems." By meeting these testing requirements and receiving certification, companies should find it easier to gain permits — a positive step for indoor reuse and large-scale commercial greywater systems.

SUMMARY OF OPTIMAL GREYWATER-FRIENDLY REGULATIONS

- Easy to follow.
- Performance-based. Guidelines outline health and safety requirements. For example, no pooling or runoff, minimize contact with greywater, keep all greywater on the property where it's generated, etc.
- Do not require a permit or fee for the safest situations; for example, single-family homes where all greywater is used for irrigation in the yard.
- Permit required only for more risky or complicated situations; for example, large flows, indoor reuse, or multi-family dwellings.
- Code is statewide with a mechanism in place to educate local regulators.

WATER UTILITY PERSPECTIVE

Susie Murray, water resource specialist with the city of Santa Rosa, California, discussed the cost-effectiveness of promoting greywater:

"From a water utility perspective, offering an incentive for greywater is cost effective. In 2012 wholesale water cost approximately $700 per acre-foot (af), while buying it back from our account holders through a greywater rebate program costs about $450 per acre-foot. Greywater also works well in combination with other conservation incentive strategies." In addition to greywater rebates, the city offers other incentives, such as for turf removal and switching to low-flow showerheads.

Ask your water district for an incentive! Water districts around the country offer financial incentives for reusing greywater. Tucson Water, in Arizona, offers a rebate for one-third the cost of a greywater system (up to $200) and offers free educational workshops and online resources (see Resources). In California, San Francisco's water department has a free manual instructing residents how to build simple systems, offers free workshops, and subsidizes parts for constructing a washing machine greywater system. Other towns, including Pasadena, Santa Rosa, Santa Cruz, Goleta, and Monterey, offer rebates for installing a system.

Greywater Pioneer

San Francisco Public Utilities Commission (SFPUC)

The SFPUC provides water to 2.6 million people in the San Francisco Bay Area, operates a combined sewer system, and treats more than 80 million gallons of wastewater a day. A leader in on-site reuse of alternate water sources, including greywater, rainwater, stormwater, blackwater, and foundation drainage, the SFPUC has developed technical assistance materials to help developers reuse water in large commercial and mixed-use residential properties.

Along with San Francisco's Non-potable Water Ordinance, the SFPUC provides financial incentives to buildings over 100,000 square feet to reuse water for irrigation and toilet flushing. Its new headquarters, located at 525 Golden Gate Avenue, in the heart of downtown San Francisco, consumes 60 percent less water than similarly sized buildings. The on-site wastewater treatment system uses a "living machine" to clean greywater and blackwater to be reused for toilet and urinal flushing, treating 5,000 gallons a day to supply 100 percent of the non-potable building water needs. The building's 25,000-gallon rainwater harvesting system provides irrigation around the building.

There are now 15 on-site water systems installed in San Francisco and 28 more in the design or construction phase, with an impressive overall water savings of 54.3 million gallons per year (this is the estimated savings).

The SFPUC's headquarters in downtown San Francisco recycles 5,000 gallons of greywater and blackwater a day to use for toilet flushing in the building.

Encouraging the reuse of nontraditional water supplies, such as greywater and blackwater, expands our options and decreases the potable water used for flushing toilets and irrigation.

I spoke with Paula Kehoe, Director of Water Resources with the SFPUC, about their sustainable water programs.

What advice do you have for other water agencies interested in promoting on-site water reuse in urban areas?

Cities are on the front line and can create pathways to encourage innovation in water use. Water agencies are working hard to increase water efficiency and to diversify their water supply. Encouraging the reuse of non-traditional water supplies, such as greywater and blackwater, expands our options and decreases the potable water used for flushing toilets and irrigation.

Water agencies can provide leadership to address concerns with these projects by engaging with local health and building departments to establish standards for local oversight that ensure public health protection. Consistent and ongoing communication with local agencies is the key to developing successful programs.

What are your goals for city water systems?

Our overall goal for San Francisco's water use is to maximize efficiency and decrease potable water consumption. One strategy is to maximize the on-site reuse of alternate water sources on multiple scales in all new developments in San Francisco, including building, block, district, and neighborhood scales.

What aspects of the program do you think can be successfully replicated in other areas?

The Non-potable Water Program (2012) coordinated three City agencies to streamline the permitting process for on-site systems; it's a great model for other municipalities. We worked with the City's Departments of Building Inspection (SFDBI) and Public Health (SFDPH) to develop a regulatory pathway to approve alternate water source projects. The SFDBI oversees construction; the SFDPH prescribes water quality criteria and oversees operation (including permitting the treatment systems); and the SFPUC provides cross connection control services, technical guidance, and financial incentives up to $500,000 per project to encourage on-site reuse.

We've amended the program twice; first to allow buildings to share non-potable water, and then to require the installation of on-site water systems for toilet flushing and irrigation in new developments meeting specified criteria (projects over 250,000 square feet). Mid-sized developments (40,000 to 250,000 square feet) aren't required to install an on-site system, but they do have to submit a water budget application to the SFPUC.

ATLAS
NITRILE
370M

PART 2

Building Your Home Greywater System

Get your gloves on and the measuring tape out — it's time to design and build your system.

By doing the preparation work in Part 1 you learned about how much greywater your home generates, your soil types and the size of infiltration areas needed, and the plants you'd like to irrigate. In Chapter 6 you read about different systems and determined which system(s) are a good match for you. In this section you'll learn everything you need to know about designing and installing a few of the most common types of greywater systems. You'll also find an overview of other types of systems — either more complicated or less common.

CHAPTER 8

Install a Laundry-to-Landscape (L2L) System

Invented by greywater pioneer Art Ludwig, the laundry-to-landscape (L2L) greywater system is one of the most popular types of systems in the U.S.

I've helped install hundreds of these in workshops and trainings and can't imagine living without one. It can be built with off-the-shelf parts, doesn't alter the household plumbing, and doesn't require a permit in many states. The system captures greywater from the drain hose of the washing machine, connects to a diverter valve so you can easily switch the system on or off, and then distributes greywater into the landscape through a main line of 1-inch irrigation tubing and ½-inch branch lines that feed specific plants. It is one of the easiest systems to construct, and is very easy to change.

This chapter will lead you through the specifics of how to design and install your own L2L system. I'll also discuss where to source materials, what maintenance is required, and how to construct the mulch basins.

IN THIS CHAPTER:

→ Design Considerations

→ Installing an L2L Irrigation System

→ Irrigation Options

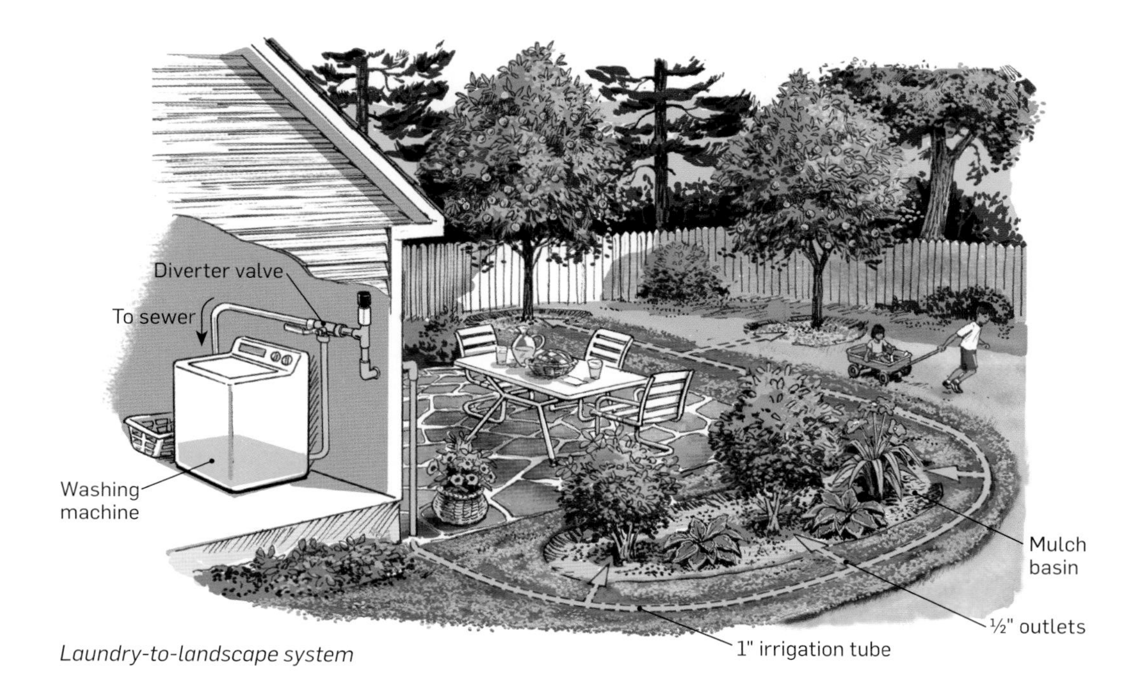

Laundry-to-landscape system

Design Considerations

To prepare for constructing the system, you'll determine where to run the pipe from the washing machine out to the landscape, what plants are easiest to irrigate, and how many plants you can irrigate based on your home's greywater production. By choosing nearby plantings or those downhill of the home, you'll design the system so the washing machine's internal pump is not overtaxed. The type of washing machine you have will impact the system: conventional top-loaders can distribute greywater to more locations than front-loading machines and most high-efficiency top-loaders.

Note: If your home generates large volumes of greywater, be sure to read Got a Lot of Greywater? Don't Dig One Big Basin (page 122) and Irrigating with Multiple Zones (page 154) for advice on appropriately distributing larger flows.

Adding Gravity-Feed?

If you're planning to install a gravity-fed branched drain system as well as an L2L system, it's wise to first plan the irrigation portion of the gravity system before deciding what to irrigate with the L2L system. The reason? It's much easier to distribute water from an L2L system than a gravity one, and you'll have more options for the L2L. See page 120 for guidance on designing a branched drain system. In general, if one area in the landscape will receive greywater from both types of systems, direct greywater from the branched drain system to irrigate the closest trees and bushes, and use the L2L system to serve slightly more distant plants or smaller plants in the vicinity.

Distribution Limitations

The washing machine pump pushes greywater through the L2L irrigation system at low pressures. Each time the machine pumps it sends all the water inside the washer drum to the landscape, then the pump turns off. Because of this, the system is limited in how many outlets it can have; if there are too many, it's impossible to get water to come out of them all. You can always have fewer outlets, but including more may require a lot of adjusting to achieve an even flow of water — if it's even possible. Here are some general guidelines on the limitations of different types of machines.

- Regular top-loading machines can distribute water up to 20 locations maximum (19 tees and one open end of the tubing).
- Water-efficient machines can distribute water up to 8 locations maximum (7 tees and one open end of the tubing) for a front-loading machine, and 10 for a top-loading machine. Note that for an ultra-efficient front-loading machine you may need to reduce to 4 outlets.

Assessing Your Site

Identify planted areas suitable for L2L irrigation (trees, bushes, larger perennials, or large annuals). Then, consider where the irrigation pipe from the washing machine can exit the house, either through the wall or crawl space. Can you exit near a suitable irrigation area? If not, are there other exit options (for example, crossing to another side of the house through the crawl space)? Also consider landscape alterations to make the installation easier: for example, planting trees or bushes in an accessible part of the yard.

HOW HIGH AND FAR CAN THE WASHING MACHINE PUMP?

The washing machine pump can send greywater directly to the landscape, but not too far away or up a hill. Overworking the machine's pump could damage it. Since replacing a washer pump is a lot cheaper and easier to do than buying and installing an effluent pump system, some people whose yards are gently sloped choose to overwork their machine's pump slightly rather than installing a pumped

WASHING MACHINE WARRANTIES

If you are purchasing a new washing machine, you may be wondering whether installing this system could affect your washing machine warranty. The answer: Perhaps, but probably not. My personal and anecdotal experience is that washer repair people don't question the greywater system or attribute any washer problems to it. I know of just one instance when the company's installer of a new washer didn't want to hook it up to the pre-installed 3-way diverter valve, claiming it would void the warranty on the machine. Others have experienced the opposite: the installer helpfully hooked up the new washer to the valve.

system. These general guidelines will help you select an appropriate part of your landscape, but remember that each situation is different; change the system if you notice problems with the machine. If your machine doesn't drain well or has other pump problems, fix them before installing an L2L system.

- **Upward-sloped yard:** Don't distribute water uphill from your washing machine, meaning the yard should not be higher in elevation than the machine. The pump in the machine is not designed for this. If your yard has a very slight upslope, send the greywater pipe to the highest irrigation point and then irrigate downslope.
- **Downward-sloped yard:** On downhill slopes the greywater distribution piping can extend as far as necessary. Don't run tubing straight down a steep slope, though; the water will rush to the bottom, making it difficult to irrigate the upper portions. Instead, snake tubing down the hill in an S, or serpentine, pattern, like a switchback trail, to slow the flow of water.
- **Flat yard:** Most machines can pump water across a flat yard for up to 50 feet without a problem. Traveling farther risks pump damage because friction in the tubing increases pressure on the pump. Elevating the machine can help give a little more oomph in the system.

Distributing water to landscapes with different slopes, using an L2L system

Subsurface or Surface Irrigation?

There's a wide range of recommendations among different state greywater codes regarding where to direct greywater — either on the surface of the landscape or delivered deep belowground. Because most of the plants' roots, as well as the soil microorganisms that clean greywater, are found in the upper 1 to 2 feet of soil, surface irrigation is usually better. (Remember, greywater must land onto mulch, never bare soil.) It's also a lot easier to observe and maintain the system when greywater outlets are aboveground; people are less likely to forget to check the outlets since they're visible, and maintenance is easier as well.

Surface irrigation isn't always possible, though. Some codes don't allow it, and often gravity-branched drain systems must have outlets below grade due to the slope of the land. Whenever you need to irrigate subsurface, try to stay as close to the surface as you can. Some states require that outlets are covered but not necessarily below grade. For example, California's code requires a 2-inch covering, including a "solid shield," so the outlet can be on the surface covered by a mulch shield.

For details, see Irrigation Options on page 113.

Subsurface irrigation into a mulch basin. A lid (not shown) covers the outlet.

Surface irrigation into a mulch basin

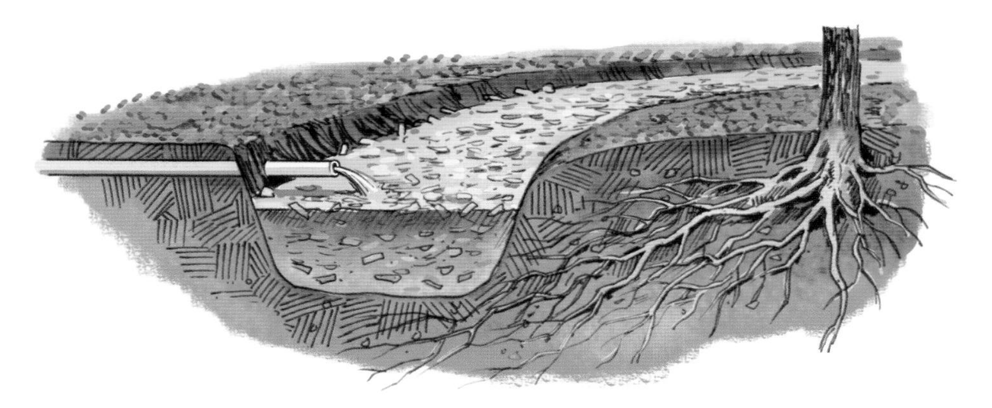

SURFACE IRRIGATION
Greywater drops through the air and lands onto mulch, where it's quickly soaked up.

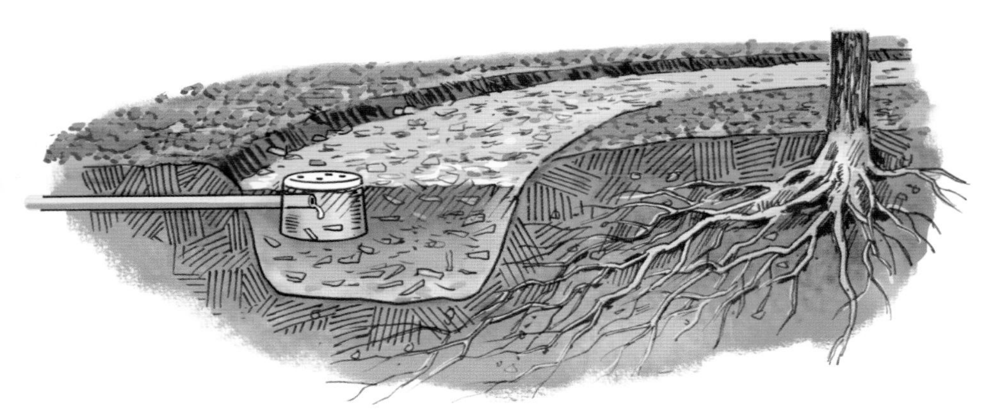

SUBSURFACE IRRIGATION
A mulch shield protects the greywater outlet and water is released below the surface of the mulch.

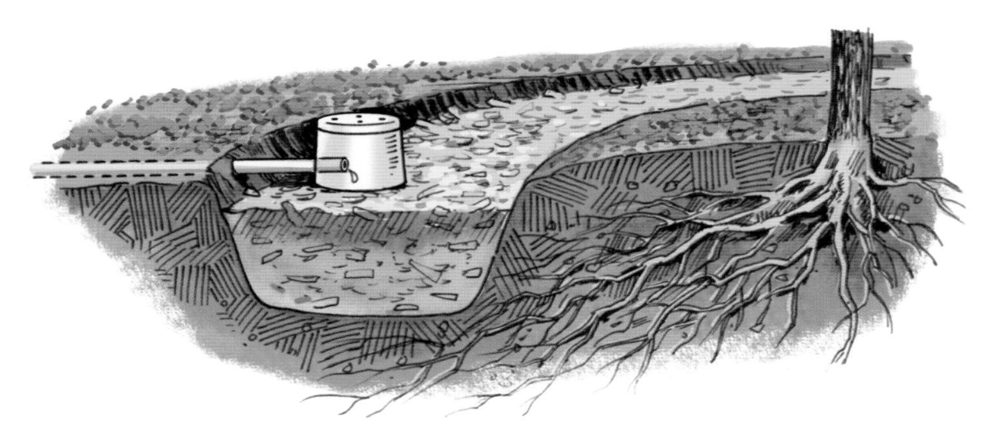

COVERED OUTLET AT SURFACE
Greywater lands onto mulch at the surface and is covered with a solid shield.

Sketching and Documenting Your System

As you're determining where you'll send the water — and before beginning construction — make a detailed diagram of the entire system, including the number of bends, length of pipe, etc. This will save you a few trips to the store when you build. Remember that the end of the main tubing can be an outlet and be located in a mulch basin near a plant to be irrigated. Alternatively, if your site requires that the end of the line is capped off, use a fail-safe overflow in the system to protect the machine's pump (see page 109).

After you've installed the system, edit this drawing as necessary to document the "as-built" system. This document, along with the pre-burial photographs, will help you find the buried tubing and outlets in the future. Add all final drawings to your Operation and Maintenance Manual (O&M) manual (page 64).

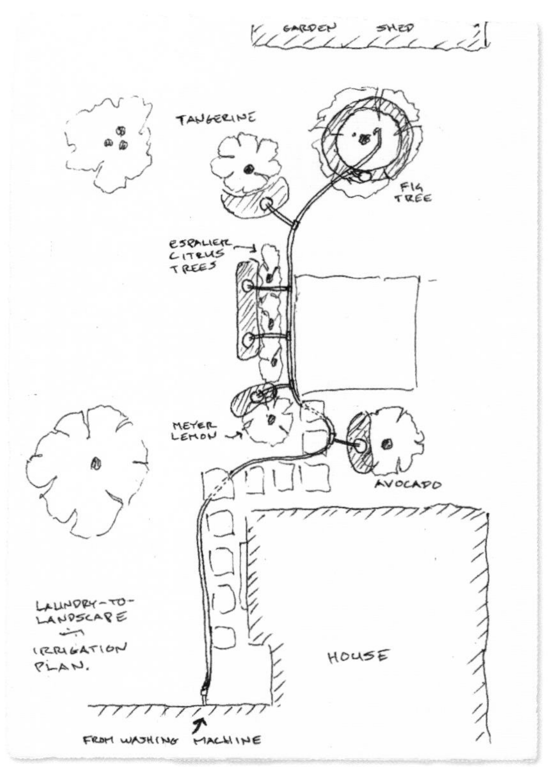

Sketch of system

KEEP THE PUMP FILTER CLEAN

Most front-loading washing machines have an internal filter to keep large particles out of the pump. Over time the filter clogs, which adds strain on the pump and can prevent it from evacuating all the water from the machine. Before installing a greywater system and anytime the machine struggles to pump out the water, clean the pump filter. Remove the front cover to reach the filter, unscrew the filter cover, and be prepared for a few gallons of water to spill out. Clean out the filter and put it back in place. If you don't have the owners manual, you can find instructions online for various models of washing machines by searching with the phrase "how to clean your pump filter," or you can consult a washing machine repair person for assistance.

Hardscape in the Way?

Greywater pipes can't run over hardscape. If you need to pass through a driveway, stairs, or a patio to run the pipe to the landscape, you can:

- Remove the hardscape.
- Go under it. If the sidewalk is just a few feet wide and there is enough room to dig on both sides, tunnel under. First dig a hole on each side of the hardscape, then start digging under it with a trowel. Connect the holes with a steel rod or other ramming tool. Use rigid 1-inch PVC pipe instead of tubing under the hardscape if you think the tubing could be damaged when you shove it under. Tape off the end of the pipe so soil doesn't get inside, then hammer the pipe through the tunnel. Or use a water-jet to tunnel under, though it makes a giant mud puddle.
- Cut a groove out of the hardscape, using a masonry blade on a circular saw, or a concrete saw. Fill the groove with concrete patching material or cover it with gravel, sand, or decorative tile.
- Run the pipe on top of the concrete, then build a small ramp cover for the pipe (so it's not a tripping hazard).
- Find an alternate route — perhaps along the house to the edge of the hardscape. This may alter your irrigation plans.
- Run the pipe snugly against a stair riser to avoid creating a tripping hazard. Use 1- or 2-hole straps to attach the pipe snugly.

Working pipes around hardscape

System Design Example

Here is an example of an L2L system for a three-person household in Austin, Texas. They wash four loads of laundry a week at 25 gallons a load and never wash more than two loads on any given day. This is the basic information they need to design the irrigation and infiltration area for their system:

- **GREYWATER PRODUCED:** 100 gallons a week (25 gallons × 4 loads).
- **DAILY MAX:** 2 loads (50 gallons); this will help determine the size of their mulch basins.
- **SOIL TYPE:** Sandy clay (coefficient 0.6; see page 43); 0.6 × 50 = 30 square feet of mulch basin
- **PLOT PLAN OF HOUSE YARD:** Washing machine on east side; plants on the south side. They can run the pipe under house to reach the south side.

When they assessed the site, they found that their existing irrigation system had three zones. One of the zones irrigated five small trees (3-foot radius from trunk to drip line), which were the most suitable plants for an L2L system (the other plants were small and spread out). They could slope the pipe downward as it traveled under the house, then run it 10 feet to reach the first tree; all the trees were within 50 feet of where the pipe traveled flat. This zone of trees seemed to be easiest to irrigate, so they checked to see if the greywater production was a good match for the trees (was it too much or too little water?). In terms of irrigation amount, each tree would

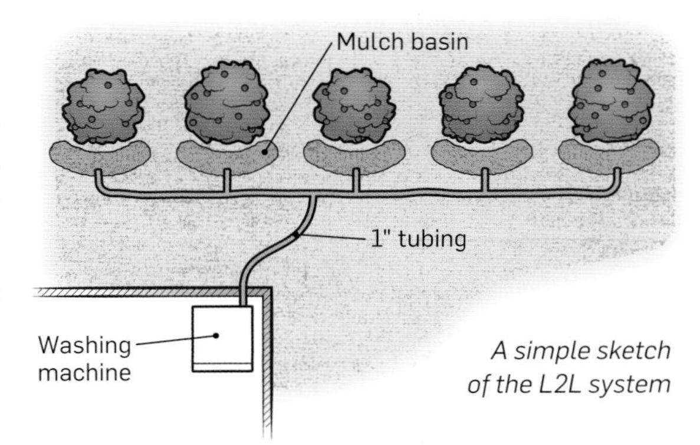

A simple sketch of the L2L system

receive 100/5 = 20 gallons per week. Austin gets some summer rain to supplement greywater during the hottest time of the year.

They calculated the plant water requirements using both methods from chapter 5, pages 50–53.

Using the "rule of thumb" method, one small tree requires:

$\pi r^2 = 3 \times 3 \times 3$ feet = 27 square feet

27 square feet × 1 (hot climate) = 27 gallons per week

Using the ET method, a tree of this size in Austin requires:

0.62 × 27 square feet × 0.5 × 8.21 (ET) = 69 gallons/month or 17 gallons/week

This demonstrates that 20 gallons a week is a good amount of irrigation water for each tree in this climate.

Next, they determined how large each mulch basin should be to infiltrate greywater into the ground. Based on the soil type and gallons per day calculated above, the system requires 30 square feet of mulch basin area. Since the water will be divided among five different trees, each tree needs basin at least 30 ÷ 5 or 6 square feet.

INSTALLING A "SECOND STANDPIPE" FOR A WASHING MACHINE

A "second standpipe" system is another option for a washing machine greywater system. It includes a secondary standpipe installed near the washing machine to drain by gravity to the landscape. To switch between the greywater and the sewer/septic systems, you manually move the washing machine's drain hose from one standpipe to the other.

Second standpipe systems are most commonly used in hot climates that have relatively high water requirements per tree, because it's harder to distribute greywater to many plants. The irrigation portion of the system is installed with a branched drain configuration system (see Installing a Branched Drain System on pages 127–141 for installation details).

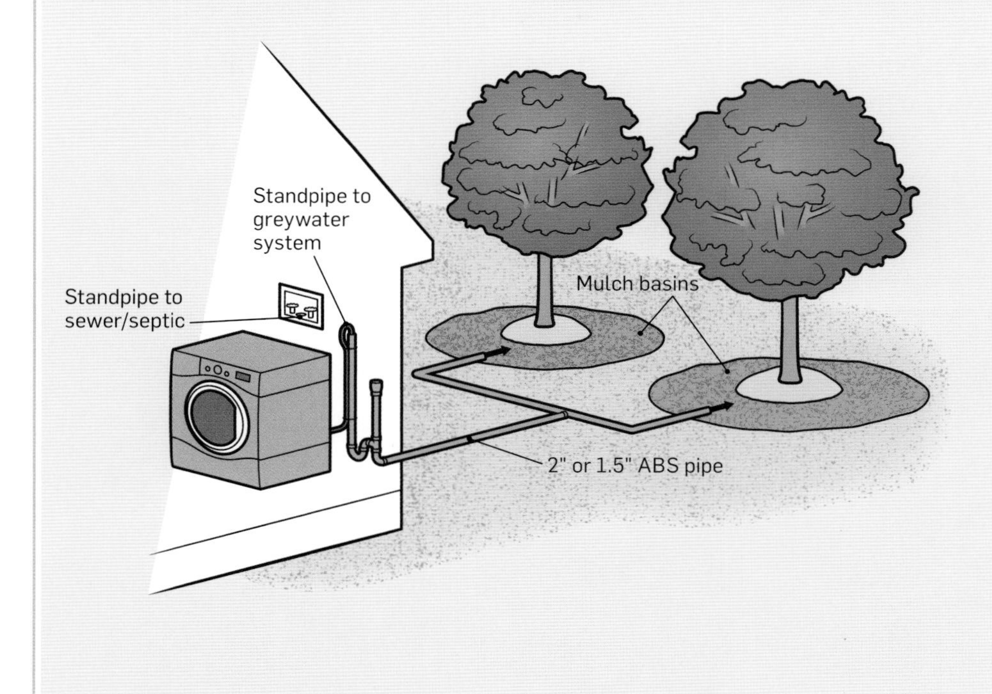

Installing an L2L IRRIGATION SYSTEM

Installing an L2L system requires very basic plumbing and landscaping skills. You'll be working with plastic pipe and tubing, which are easy to cut and assemble. In general, the most difficult aspect of the installation is drilling the exit hole for the pipe. If drilling through the wall or floor of your home feels daunting, get help from a knowledgeable friend or hire a handy-person to help you. Most systems take a day or two to install, though with help can go more quickly. Make sure to source the materials for your system a few weeks before you plan to install it. Depending on what your local hardware and irrigation stores stock, you may need to mail-order a few of the parts; see Resources for online retailers. You can also build a PVC-free system using Blu-Lock or Eco-Lock fittings (see Resources).

See Plumbing Basics for Greywater Installation (page 178) for details on parts and basic techniques.

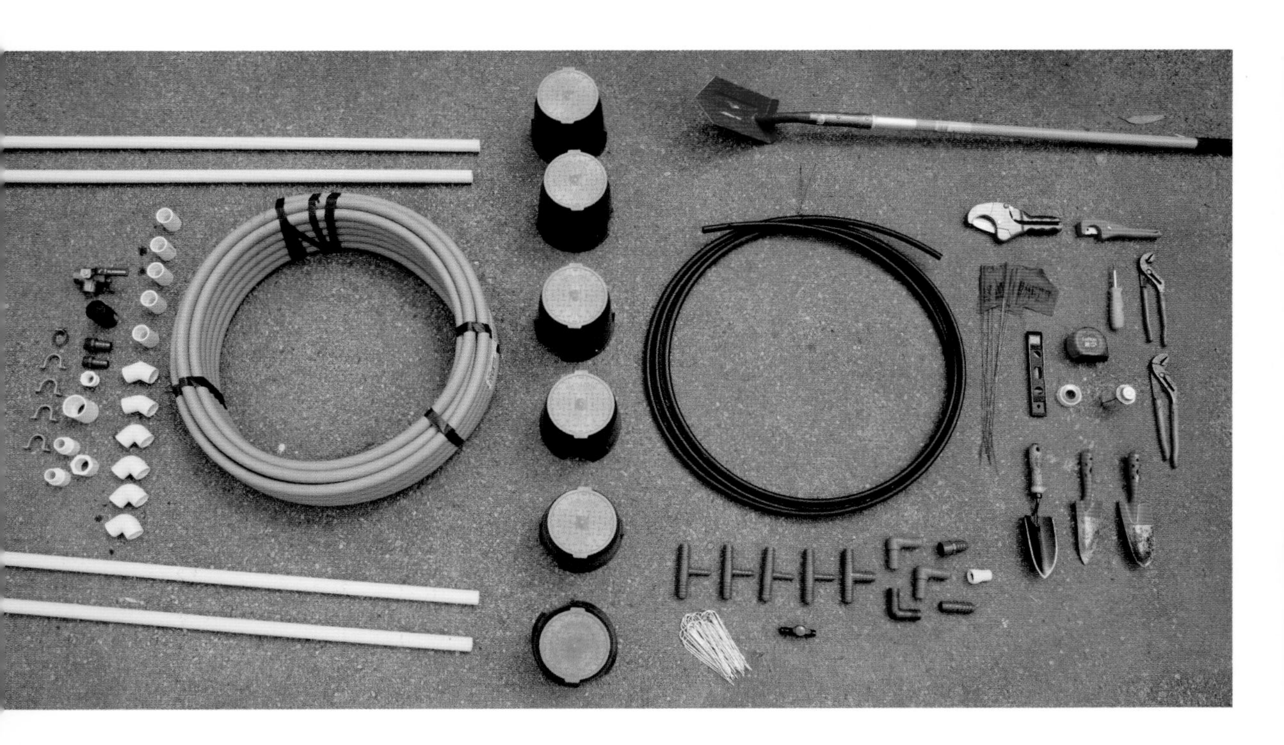

Materials

INTERIOR PLUMBING

- Piece of wire
- Pipe thread tape
- Two 1" PVC male adapters (slip x MPT)
- One 1" PVC male × barbed adapter (or 1" male × ¾" barbed adapter if your washer hose is smaller; measure the inner diameter of the end of your washing machine drain hose to determine which size you need. Or use a ¾" barbed adapter screwed into a 1" × ¾" bushing instead)
- One brass threaded 3-way valve (L-type, full-port valve; see step 2)
- One 1" hose clamp
- Two 1¼" 2-hole straps with screws
- 1" schedule 40 PVC pipe (as needed)
- 1" PVC 90-degree and 45-degree elbows (as needed)
- One air admittance valve (AAV) or auto-vent
- One 1" × 1½" PVC reducer bushing
- One 1½" PVC female adapter (slip × FPT)
- One 1" PVC tee
- PVC glue (solvent cement; use PVC primer, if recommended)
- 1" PVC couplings (as needed)
- 1" 2-hole strap with screws (as needed)

LANDSCAPE IRRIGATION

- One 1" barbed × slip (or barb × insert) transition coupling
- 1" HDPE tubing (as needed)
- ½" PE tubing (as needed)
- 1" × ½" barbed reducing tee (one for each "irrigation point," or branch from the main line to each mulch basin)
- 1" × 1" × 1" barbed tee (quantity as needed; one for each branch from the main line)
- Large garden staples (for ¾" or 1" tubing; use one every few feet)
- ½" green- or purple-backed ball valves (full port ball valves; quantity as needed, usually 2 to 4)
- Exterior house paint
- Construction-grade sealant (see step 15)
- Optional, for flushing the system with a garden hose: either one ½" slip x ¾" female hose thread (this fitting inserts into a 1" barbed × slip fitting); or one ¾" slip × ¾" female hose thread (this fitting inserts into 1" pipe); or one 1" PVC slip union and one brass female hose thread × ¾" male hose thread fitting. (See step 12)

Tools

- Drill and ¼" pilot bit (long enough to go through your wall or floor)
- 1½" hole saw (optional: masonry bits for stucco)
- Flathead screwdriver
- Two tongue-and-groove pliers (or two large wrenches)
- Hex-head (nut) driver
- Permanent marker
- PVC pipe cutter or handsaw
- Level
- Shovel
- Mattock or pickax
- Narrow trenching shovel
- Tubing cutters or large clippers
- Hammer
- Thermos (full of hot water) and a cup
- Painting supplies and caulking gun

FINDING PARTS TO BUILD AN L2L SYSTEM

Gathering parts for your system will take you to several stores (irrigation, plumbing, and hardware) or online retailers (see Resources).

TUBING. Find 1" tubing, either high-density polyethylene (HDPE) or polyethylene (PE, or "poly"). HDPE is stronger and less likely to kink than PE. Irrigation stores sell several types of HDPE tubing, all suitable for greywater. Don't use tubing that is visibly kinked or seems easy to kink (try bending it to see). In urban areas the only 1" option typically is *Eco-Lock* from Home Depot or *Blu-Lock*; find a distributor on the manufacturer's website (Hydro-Rain).

3-WAY VALVE. Brass 3-way valves (L type), used in hydronic heating, often are hard to find locally. Larger plumbing supply stores carry them, and sometimes general building supply stores or irrigation stores will order them. Get a *full port* valve, available from brands such as Legend Valve, Webstone, and Red-White. The inside of a full port valve is not reduced in size, whereas *regular port* valves are a pipe size smaller inside, which creates more friction in the system. Banjo makes a suitable full port 1" plastic valve.

AIR ADMITTANCE VALVE (AAV). Use either an official AAV ($20) or a lower-cost auto-vent or in-line vent ($4). Both are available from plumbing suppliers and hardware stores. Note that the cheaper one isn't allowed in standard plumbing applications but often is okay to use in an L2L system because it's connected only to the irrigation system, not the household plumbing.

MULCH. Use chunky wood chips (1 to 1½") from a local tree trimmer. One-half of a cubic yard is usually enough. Or, buy wood chips from a landscaping store (loose or bagged).

1" BARBED FITTINGS. Find these at irrigation stores or online (for example, dripworks.com).

PVC-FREE PARTS. Source Eco-Lock or Blu-Lock from a local irrigation store or order from Home Depot. Use a 1½" × 1" poly reducing coupling made by Banjo to connect the AAV.

PVC fittings to connect to 3-way valve

System using Blu-Lock fittings

INSTALLING THE SYSTEM

1. MAKE AN EXIT HOLE FOR THE PIPE.

To exit the house you need a 1½" hole, either in the wall or floor. Identify the best location for the hole. Exit through the wall if you don't have a basement or crawl space, the area you plan to irrigate is adjacent to the wall, or the washing machine is on the second story and there is living space below the floor. Exit through the floor if you have an accessible basement or crawl space and can run the pipe out through a screened vent or other opening (you won't have to drill a hole in an exterior wall or foundation wall).

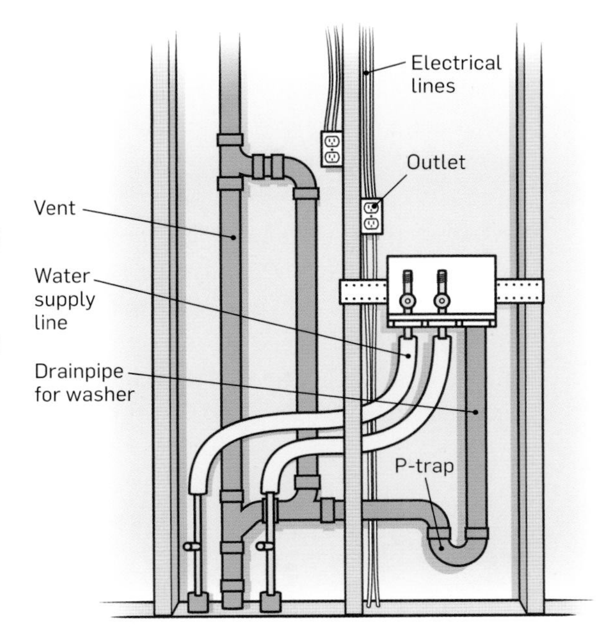

Watch out for plumbing and electrical inside the wall when you drill.

Before you drill, look for hidden obstacles in the wall, such as electrical wires, pipes, or studs. Drill a ¼" pilot hole with a long, thin drill bit (called a pilot bit) that can pass through the entire wall or floor. Use a masonry pilot bit for stucco walls. Drill carefully; stop if you hit anything in the wall. If necessary, try different spots until you exit freely. Insert a piece of wire into the hole and wiggle it around to check for nearby pipes. Check the exit hole from the outside to see if it's a good location, for example, not right next to the window frame. Since the pilot hole represents the center of the 1½" hole, make sure there's at least ¾" clearance around all sides of the pilot hole.

Step 1. Drilling a hole through the floor with a hole saw

When you're confident you're drilling in the proper location, use the pilot hole as a guide and drill the exit hole with a 1½" hole saw. To make a clean-looking hole on both sides of the wall, drill from both the inside and the outside. When drilling through several wall layers, the hole saw will get filled up and you must remove the "cookie" that forms inside the bit (using a screwdriver) before you're able to drill all the way through. Be sure to drill either straight out, or at a slight downslope, so the pipe will gently slope downwards, directing any rain that lands on the pipe outside away from the building.

If your wall's exterior siding is wood or vinyl, use a regular hole saw bit for the hole. If the siding is stucco or fiber cement siding, you can use a special masonry hole saw bit, but these are expensive. Alternatively, trace the outside of a scrap piece of 1" PVC pipe on the wall with a marker, then drill around the hole with a smaller masonry bit. Chip out the circle with a hammer and chisel, then snip the lath or wire behind with metal snips (for stucco). After the stucco or fiber cement is removed, drill the rest of the way using a standard hole saw.

Note that some building codes require exit holes to be "fire rated" if they are within a certain distance from another building; for example, 5 feet. If other homes are very close to yours, check with your permitting agency about fire rating requirements.

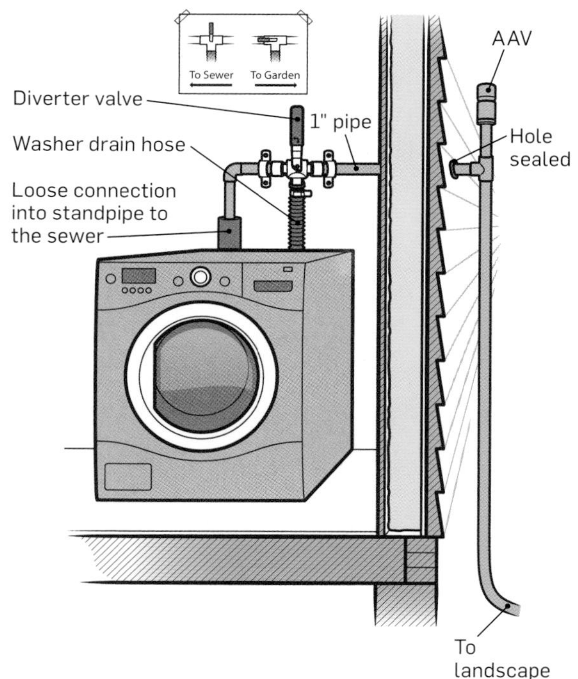

L2L indoor plumbing: Washing machine connected to diverter valve; water flows either to landscape or to sewer/septic.

2. PREPARE THE 3-WAY VALVE.

Wrap pipe thread tape clockwise around the threads of the two 1" male (non-barbed) adapters (clockwise as you face the open end of the threaded side), wrapping tightly three times around. Screw the adapters into the right and left sides (opposite ends) of the 3-way valve. Start gently by hand to avoid cross-threading. (If you cross-thread, the metal threads will strip the plastic threads and will leak.) The fitting should turn easily at first, then get harder. Do the same with the 1" male × barbed adapter (or 1" male × ¾" barbed adapter for smaller washer hoses) in the middle outlet of the valve. Tighten the fittings snugly with tongue-and-groove pliers. Don't overtighten, or the fitting could crack.

Place the hose clamp over the washing machine drain hose. Then push the hose over the barbed fitting on the adapter. Tighten the hose clamp over the hose and barbed fitting, using a flathead screwdriver or hex-head driver.

Note: Some hoses have a hard end portion that can be cut off. Others have a rigid U-shaped piece of plastic attached to the hose that can be removed. If the hose is too short or otherwise unsuitable, replace it (it's easy to do). If it's hard to fit the hose over the barb, soften the hose by

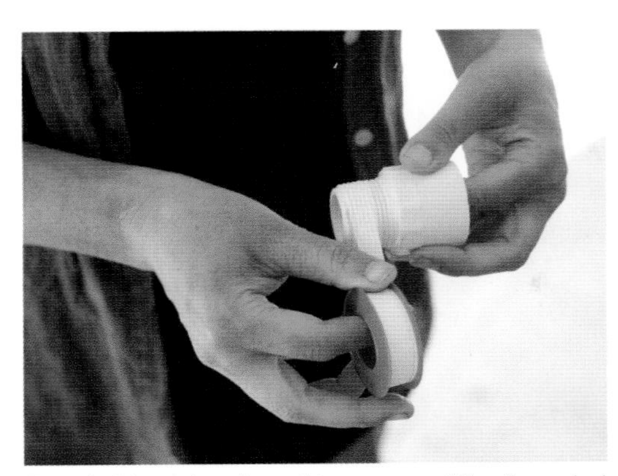

Step 2: Wrap pipe thread tape around the threaded fittings.

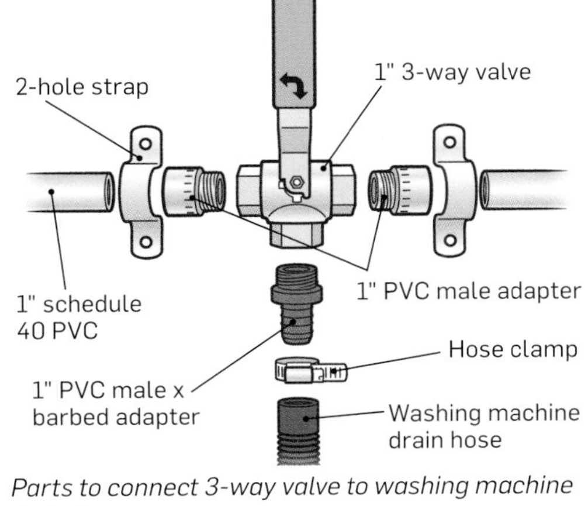

Parts to connect 3-way valve to washing machine drain hose

First screw the threaded fittings into the valve by hand, being careful not to cross-thread. Then tighten them with tongue-and-groove pliers.

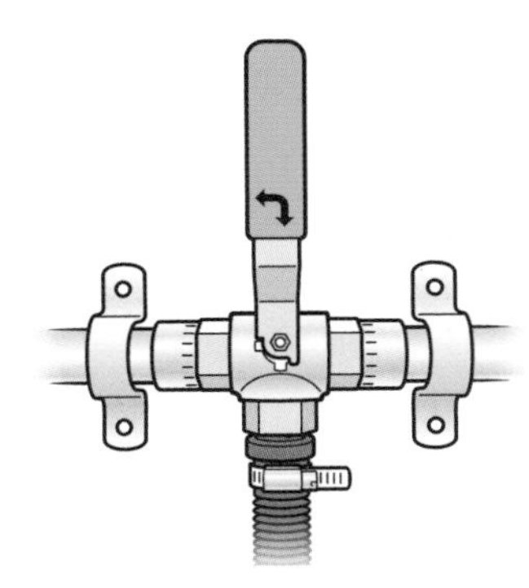

3-way parts together. Pipe thread tape used on threads and PVC glue on pipe joints create a watertight seal.

Attach the washer hose to the valve with a hose clamp.

heating it with a hair dryer or sticking it into a cup of hot water; when hot, the hose should fit on. Check this connection for leaks after you finish the system.

3. MOUNT THE VALVE.

Decide where to mount the 3-way valve. The valve must be above the **flood rim** of the washing machine (the highest place water could fill) and have enough clearance so you can easily turn the handle. The valve must be strapped to a stud or other piece of wood mounted behind. (If there is no stud, attach a piece of wood behind the valve location.) Mount the valve with 1¼" 2-hole straps (if you strap over the pipe, use 1" straps), using a level to keep valve straight. The valve can be oriented in any direction (such as sideways or upside-down), but the washer hose must still connect to the middle port. Note: If you mount the valve sideways or upside-down, use a 90-degree elbow to run the pipe horizontal before the tee and AAV assembly.

4. PLUMB TO AND FROM THE 3-WAY VALVE.

Next, plumb the most direct route from one side of the valve back to the sewer/septic (where the hose previously drained). At this stage, you're just dry-fitting the parts. If your hose previously drained into a utility sink, direct the PVC pipe into the sink in an out-of-the-way location. If your hose drained into a standpipe, plumb the PVC pipe so it extends a few inches into the standpipe. The standpipe should be larger than the PVC pipe so that there is room around the sides for airflow. If your existing standpipe is undersized, consider upgrading it.

Plumb the other side of the valve to the exit hole to the outdoors. Try to maintain a

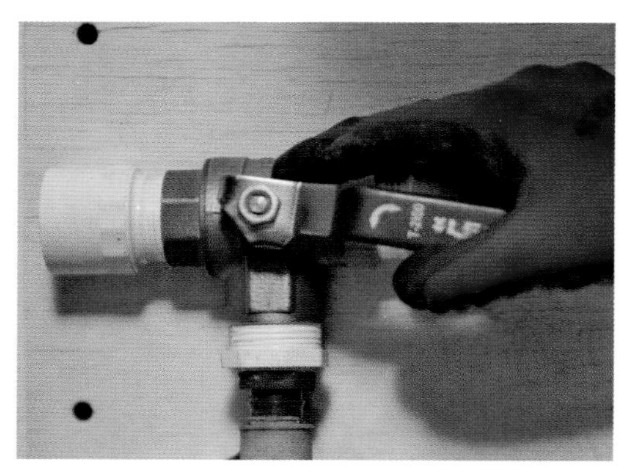

Step 3: Decide where to mount the valve. Attach a piece of wood to fasten the strap to.

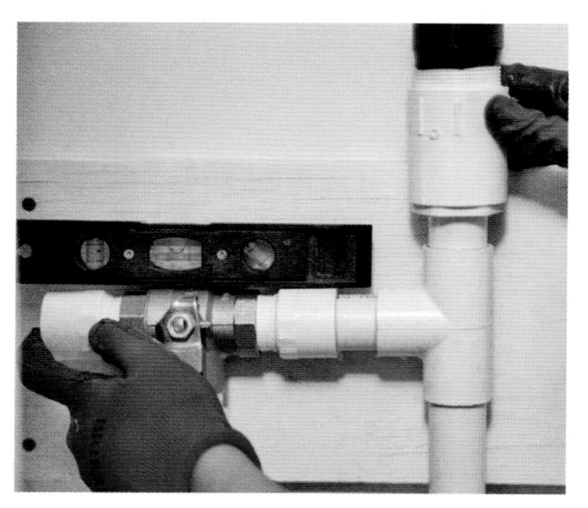

Mount the diverter valve to the wall.

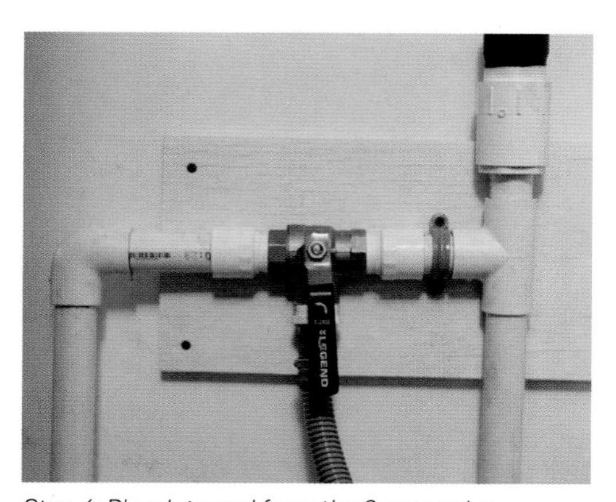

Step 4: Plumb to and from the 3-way valve.

Optional: Use a second washer hose or a flexible 1" tube to connect from the valve to the standpipe.

slight downward slope to avoid having extra water drain back into the machine. Use 1" schedule 40 PVC pipe and either 90- or 45-degree elbow fittings. Measure, cut, and dry-fit (without glue) the PVC together to get an idea where the pipe will go. With glue, the pipe will slide farther into the fitting, up to the lip on the interior; take this into account when measuring so you don't end up short. Leave several inches of pipe sticking out of the 1½" hole (outside the building or below the floor).

Locate the air admittance valve (AAV), or autovent (see below). This must be at the highest point of the system. Position the AAV at least 6" above the flood rim of the washing machine and in a visible and accessible location.

If you pipe through the floor or drop down to exit lower than the 3-way valve, locate the vent inside the laundry room. If the pipe travels

3 LOCATIONS FOR AAV

AAV inside, above where the pipe drops down through the floor

AAV outside, when the pipe exits the building at a high point

AAV inside, before the pipe drops to exit at floor level

directly out of the wall, at the height of the 3-way valve, the AAV can be outside. The valve may need protection in freezing climates; confirm that the specific valve you use is rated for outdoor use, or install it inside. In any case, the tee and AAV assembly will be located above the point where you drop the pipe down (either through the floor or, if outside the house, down to the soil).

Note: If the plumbing to the sewer/septic feels tricky (i.e., a lot of tight turns), use a second washer hose or 1" flexible tubing instead of rigid pipe. Attach it to the valve via a second male pipe thread × barbed fitting and slip it into the standpipe instead of plumbing with PVC. You can buy a washer hose at most hardware stores.

Optional: Add a union fitting into the pipe in any location where you may want to easily disconnect it — for example, if you plan to paint the wall behind the valve and want to easily disconnect and reconnect the system, or to send a plumber's snake down the standpipe to clean out a clog.

5. ASSEMBLE THE AAV.

Glue the 1" × 1½" reducer bushing into the slip side of the 1½" female adapter. Take care that the glue doesn't drip onto the threads on the other side of the adapter. Wrap pipe thread tape on the threads of the AAV, screw it into the threaded side of the female adapter, and hand-tighten the valve. Cut a 2" to 3" piece of 1" PVC pipe. Glue one end of the pipe into the 1" side of the bushing and the other end into the top of the tee.

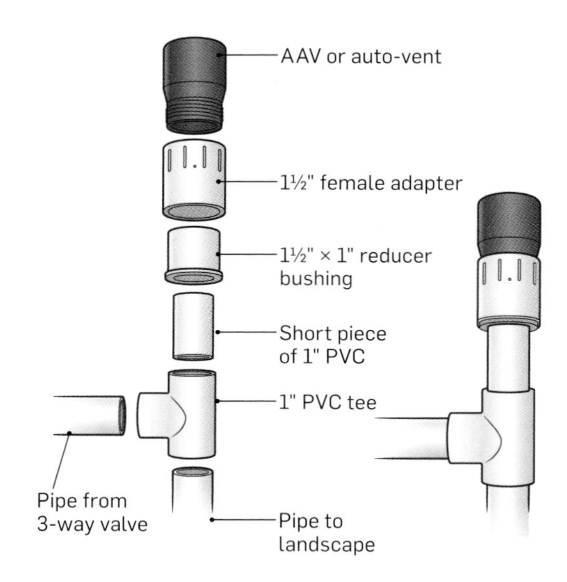

AAV assembly, separated and together

6. COMPLETE THE INTERIOR PIPE RUNS.

Now you're ready to glue the interior PVC piping. Before gluing, number and mark the fittings and pipes (with a permanent marker) so you'll know what goes where after you disassemble the parts to apply the glue. Plan the best order to glue and avoid "gluing yourself into a corner." Some of the glued joints allow for more flexibility than others; for example, the last two pieces in an assembly need space to be separated and pushed together.

7. COMPLETE THE REMAINING PVC PIPE RUN.

Depending on your site, at this point you may have a short section of PVC remaining to reach the landscape, or you may need to pipe under the house through the crawl space before reaching the landscape. If the AAV is located outside, attach it to the pipe sticking through the house wall, and pipe down to the landscape. If the AAV is inside, connect it to the pipe extending through the floor and continue to the landscape. Try to maintain a downward slope whenever possible, and minimize fittings to reduce friction in the pipe. In the crawl space, strap the pipe to floor joists or available beams every 5 feet or so. Use 1" couplings as needed to connect long runs of pipe, and glue all joints.

Outside the house, use the 1" barbed × slip adapter (or barbed × insert) to transition between PVC and the irrigation tubing. The tubing will slip over the barbed side of the fitting. This often is the most convenient place to disconnect the irrigation portion of the system to flush the line. Either use a union so it's easy to detach, or simply leave this connection unglued (it usually doesn't leak, but if a small leak would be problematic don't use this method).

Step 7: Run the PVC pipe outside.

Transitioning from PVC to HDPE tubing using a barbed x insert adapter (the gray fitting)

8. PREPARE THE LANDSCAPE: DIG THE MULCH BASINS.

To irrigate trees, bushes, shrubs, and larger annuals, dig a mulch basin in the drip line of the plant. See How to Dig a Mulch Basin (page 113) for details. If you are irrigating vegetables or herbs in a garden bed, see Irrigating Raised Beds (page 118) for advice on preparing the soil. You'll either dig very small mulch pits or cover the soil with straw and irrigate onto the straw.

Step 8: Circular mulch basin around fig tree

Trench-shaped mulch basin in front of a row of citrus trees

Level the bottom of the basin.

9. TRENCH TO EACH BASIN.

In most sites, you'll dig a trench for the tubing and bury it. Sometimes you can run the tubing on top of the soil in shaded locations; for example, along the edge of a fence, under a deck, or other out-of-the-way locations. Tubing can be in the sun, but it will last longer if it's not. You may also leave it above ground once it's inside garden beds.

Plan the run of irrigation tubing from the PVC pipe to all the mulch basins. Take the most direct route you can, with no sharp turns. Never kink or squish the tubing, which restricts the flow of water and could damage the washing machine pump. The tubing should run parallel to the basins but a few inches offset from each basin so when you install the tee the ½" tube will enter the basin (see Tubing Tips on page 102).

If there are any areas where the tubing may become damaged, such as under a swing set, transition back to rigid PVC in those locations.

Note: Avoid using 90-degree elbows for turns. Instead, make a wider-radius turn with the tubing. The tight turn in a barbed 90-degree elbow restricts the flow of water and creates a place for debris to get stuck.

Trenches should be about 4" deep and level or sloping downward. If the tubing runs up and down, it will be harder to distribute water evenly to the plants.

Step 9: Trench for tubing

Make wide-radius turns with the tubing to prevent kinks.

10. INSTALL THE MAIN LINE AND GREYWATER OUTLETS.

Roll the 1" HDPE main-line tubing to all the mulch basins (see Tips for Working with Irrigation Tubing on page 183). Make sure the main-line tubing isn't inside of the basin; it should be a few inches offset, on top of solid earth. At each irrigation point cut the tubing and insert a 1" × ½" barbed reducing tee. Dig a small trench into the basin and attach a short piece of ½" tubing from each tee and into the center of each mulch basin.

Optional: If your site requires the 1" main-line tubing to reach multiple areas, use 1" × 1" × 1" barbed tees as needed. Run a 1" line anytime you would need more than a few feet of ½" tubing to reach the plants, unless you have a very water-efficient machine, in which case it will be easier to distribute the flow by minimizing the 1" size tubing and using longer ½" runs to reach the basins.

You may discover that it's much more convenient to run tubing inside of a basin; for example, when irrigating along a narrow strip of land that isn't wide enough for the basin and tubing. In this situation, run tubing inside the outer edge of the basin and securely stake it against the side wall of the basin, as high up as possible, as shown at right. Never run the tubing in the middle of your basin, because it can be damaged or wiggled by people stepping on it.

Stake the tubing down with long garden staples so it doesn't wiggle out of the trench. You will bury the tubing later, after you've tested the system.

TUBING TIPS

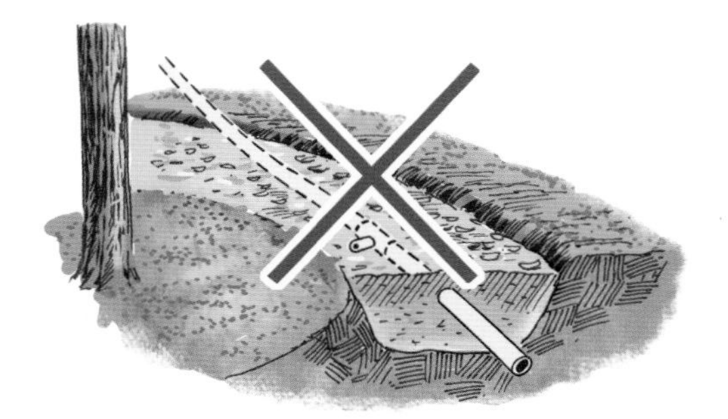

Avoid running tubing inside the basin.

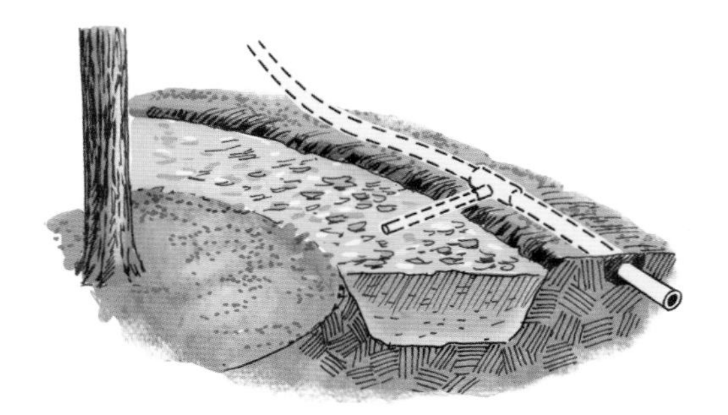

Whenever possible keep tubing buried on top of solid ground next to the basin.

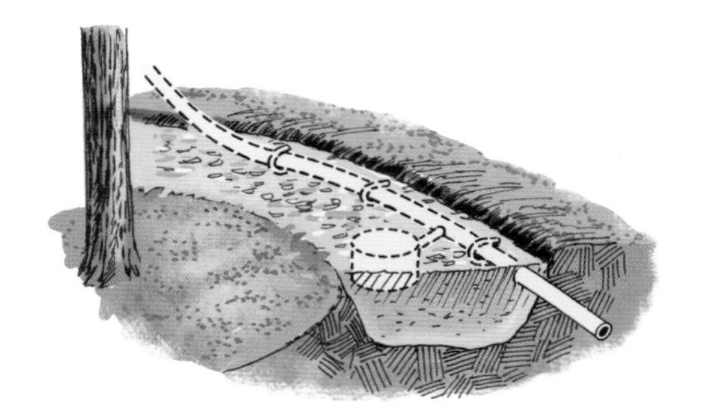

If tubing is inside the basin, stake it securely to the upper side of the basin wall.

Step 10: Lay tubing in the trench to each basin. Use garden staples to secure tubing.

Run tubing under larger roots.

Cut tubing to insert tees.

Dip tubing into hot water to soften plastic so it slides with ease over barbed tee.

Push barbed tee into tubing.

Attach ½" tubing to barbed reducing tee and run it into the center of each mulch basin.

11. PREPARE THE GREYWATER OUTLETS AND ADD MULCH.

FOR SURFACE IRRIGATION: Add mulch to fill the basins. Place a flat stone or brick under the greywater outlet to help locate it in the future and prevent it from being inadvertently covered with mulch. You'll need to be able to find the outlets easily for annual maintenance.

Greywater flows onto mulch and is quickly absorbed.

FOR SUBSURFACE IRRIGATION: Fill each basin partway with mulch, then add a "mulch shield" to the end of each outlet, including the end of the main line. See How to Make a Mulch Shield on page 115. Fill the basin the rest of the way with mulch.

Use a section of 4" drain pipe as a mulch shield.

After basin is filled with mulch, cover finished 4" drain pipe mulch shield with a paver.

Adapt an irrigation valve box for a strong mulch shield.

Fill basin just below the top of the mulch shield. After tubing is buried, the top of the mulch shield will be the only visible part of the system.

Fill the basins with mulch.

Multiple outlets in a trench-shaped basin

Finished basin with mulch shield

MARK THE GREYWATER OUTLETS. This is where you'll do the annual maintenance to ensure the outlet doesn't clog and greywater flows freely onto the mulch.

12. ADD A CONNECTION TO FLUSH THE SYSTEM (OPTIONAL).

You may include a garden-hose connection to test or flush the system. If you do this, make sure that it's not possible to connect a garden hose without fully disconnecting the greywater pipe coming from the washer. That would be considered a cross-connection and is not allowed. There are two easy ways to temporarily connect a garden hose: 1) pull apart the joint where the tubing connects to the PVC adapter and insert a fitting that connects to a garden hose (see below), or 2) install a union with some adapters to easily disconnect the system. My favorite method is to leave the barbed x insert adapter unglued where it connects to the PVC elbow, pull it out, and insert a ½" PVC slip × female hose thread fitting, to which you can attach a garden hose. (A ¾" slip × female hose thread also works and will fit into any PVC fitting.)

To use a union, open the union and insert a brass female hose thread × ¾" male hose thread (MHT) fitting into the side going to the garden. Do this by firmly pushing the brass threads into the plastic (see image) and turning— the metal will bite into the plastic enough to hold the fitting in place. Then, connect the hose and flush the line.

TWO METHODS FOR CONNECTING A GARDEN HOSE TO FLUSH IRRIGATION TUBING

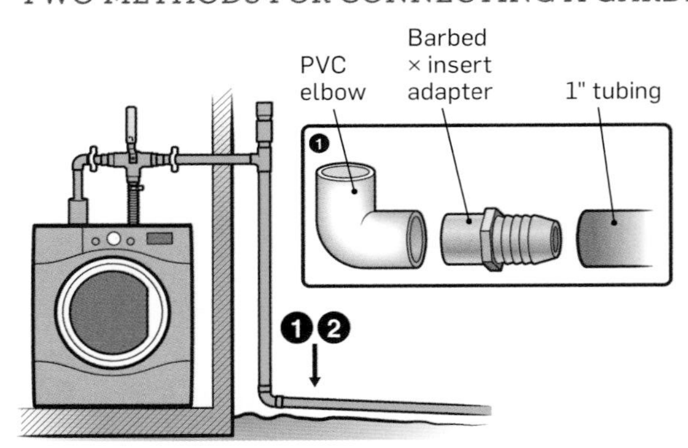

❶ *Leave connection unglued between PVC elbow and barbed × insert adapter. To flush system, pull fittings apart and temporarily insert a ½" pipe to a ¾" female hose fitting into the insert side of the adapter. Attach garden hose to other side.*

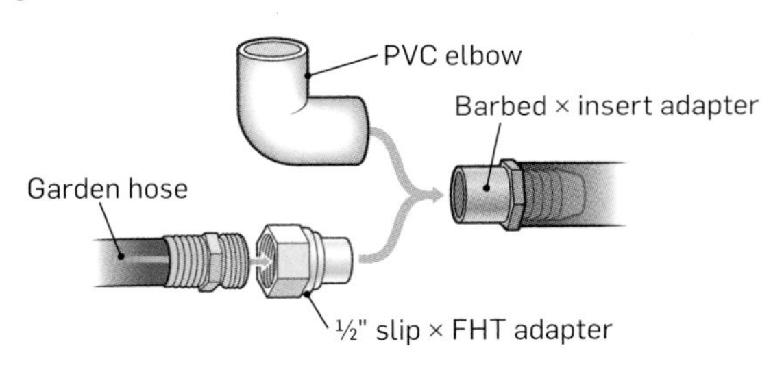

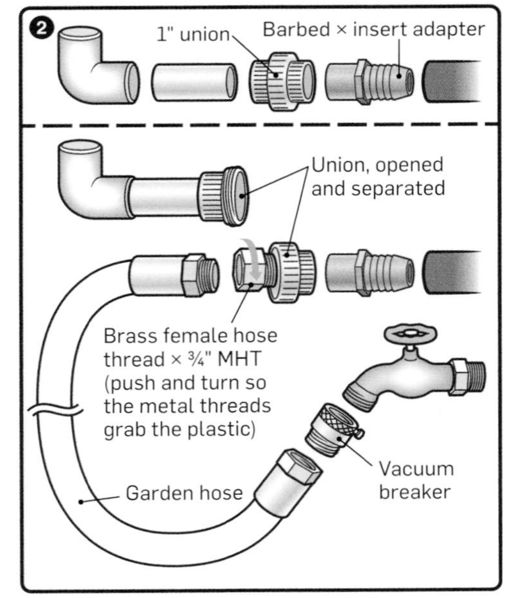

❷ *Install a union so the system can be easily disconnected. To flush the line, separate the union, use a brass hose fitting and temporarily connect a garden hose to the irrigation tubing (for flushing system).*

13. TEST AND TUNE THE SYSTEM.

Now you'll check to see if water comes out evenly from all the outlets. The first step is to get water running through the system, and there are two ways to do this. One is to run a washer cycle (or two or more), which works well with a conventional top-loading machine or in a small system with a water-efficient machine. The other method is to temporarily insert a garden hose (see step 12), and turn on the hose to medium flow. If you use this method, be sure to check the flow again with only the machine running to adjust for differences in pressure and water quantity.

Note: The first few gallons of water will fill the lines and wet the plastic tubing, and water may not come out for a few moments (possibly not until the second cycle of the machine).

If your yard is flat, you'll probably have more water coming out the first few outlets than the last ones. To tune, or balance, the system, first adjust the angles of the tees. Orient them slightly downward to get more water out of the ½" branch, or upward to slow the flow. You can also spin the ½" tubing so that the curve increases or decreases the flow.

If the flow is still uneven, add a ½" ball valve to the first outlet and close the valve slightly to reduce the flow, allowing more water to continue to other outlets. Be sure to use "green" or "purple back" ball valves. These are full-port valves with no restriction inside the valve; regular ball valves restrict the flow inside and will clog up.

Add valves until water comes out evenly, but only use them as needed. Do not add valves to all the outlets! They are clogging and

Step 13. Insert a garden hose adapter to send water through system for testing.

Connect a garden hose to the adapter.

Tune the system by adjusting the angle of the tee.

maintenance points. If you have more than a couple of valves, you probably need to redesign the main line instead of adding more valves. If your system has multiple 1" main lines and you are getting too much water in one line, consider reducing the diameter of the tubing in one of the lines, (for example, to ½") to reduce the flow.

After you've tuned the system, take photographs before burying any tubing. Put them in your O&M manual (see page 64) so you know where to avoid digging in the future.

Bury the tubing as needed. Double-check each mulch shield (if you have them) to make sure there are several inches of air space between the outlet and the surface of the mulch.

Warning: Always leave an end of the 1" main line open. Never use a valve or plug at the end to stop water flow unless there is an alternate open 1" end (see Overflow Options, opposite).

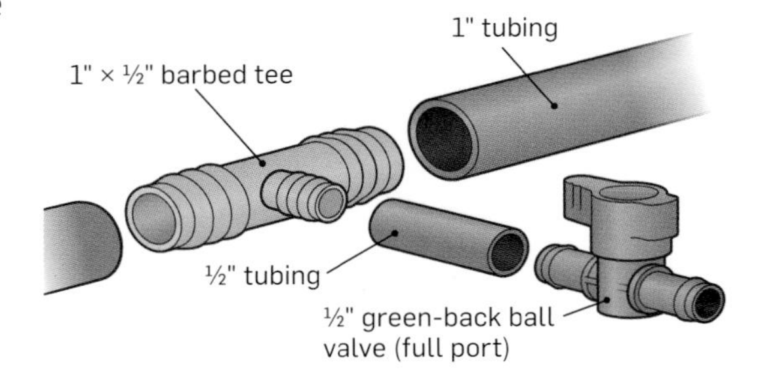

Green-back ball valve installed on outlet line to control flow

A ½" ball valve installed at the end of a ½" tubing outlet

Photograph system before burying tubing.

If you used a garden hose to test the system, disconnect it and reconnect the irrigation tubing to the PVC pipe. Run a load of laundry and observe how water flows through the system. You may need to make a few more adjustments to get water flowing out all outlets evenly.

Another view of system before burial

OVERFLOW OPTIONS

Restricting the main line can back up the washing machine or damage its pump if outlets clog over time. The exception to this rule is when a system has multiple 1" lines. For example, if you used a 1" × 1" × 1" tee and have two different main lines, one can be restricted, as long as the other one stays fully open. You can locate the 1" open end anywhere along the system (use a 1" × 1" × 1" tee), and even elevate it so that water doesn't flow out of it under typical conditions.

ONE OPTION FOR END OF MAIN LINE
Open 1" line in a mulch shield.

SECOND OPTION FOR OPEN END OF 1" LINE
Fail-safe overflow. Extend a 1" line off a 1" barbed tee in a discreet location (looking like a snorkel) to provide a fail-safe exit for the water in case the outlets clog over time. Add a 1" barbed 90° elbow on the top to prevent debris from falling into the line.

Using a fail-safe overflow allows you to cap or restrict the end of the mainline. Locate the overflow against the trunk of a tree, in a corner of a porch, or other out-of-the-way location. Water should not exit the overflow under normal conditions.

THREE WAYS TO RELEASE GREYWATER INTO THE MULCH SHIELD: *barbed reducing tee (top), ½" tubing (middle), and ½" full-port ball valve (bottom). All outlets should release greywater a few inches above the mulch layer to prevent roots from growing into the system and clogging it.*

14. CHECK FOR LEAKS.

With the water still flowing through the greywater system, check all of your glued joints in the house (on both the greywater and sewer/septic side). Carefully check the connection from the washing machine hose to the 3-way valve. If it leaks, tighten the hose clamp or add a second clamp. If it's still leaking, disconnect the hose and use a short piece of vinyl tubing to create a bridge. Clamp the tubing to the washing machine drain hose and to the barbed fitting.

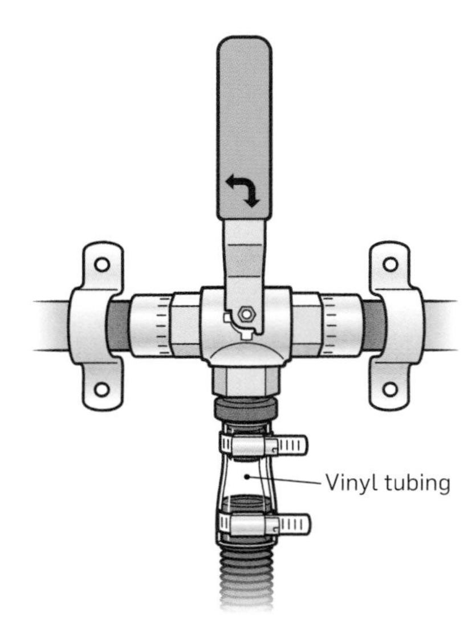

Bridge of vinyl tubing between the washer drain hose and 3-way valve to stop a leaky connection.

15. **PAINT THE PIPE AND LABEL THE SYSTEM.**

Paint any exposed exterior PVC pipe with exterior latex house paint to protect it from sunlight, which makes bare pipe brittle over time. Seal any holes in the wall or floor with a construction-grade exterior sealant that is waterproof and adhesive to protect your home from water damage. Sikaflex is a common brand. Regular silicone caulk can peel away over time.

Label the 3-way valve. Even though the valve comes pre-labeled with a small arrow indicating the direction of flow, it's not always obvious where the water is going. Make a clear label, or take a photo of the valve with the handle in each position and label the photos. Prominently post the label or photos near the valve.

Note: Many codes require aboveground pipes to be labeled with "Caution: Non-potable Water, Do Not Drink" or similar wording placed every 5 feet (see page 74).

Put your O&M manual in a visible place; for example, inside a plastic sleeve taped to the washing machine.

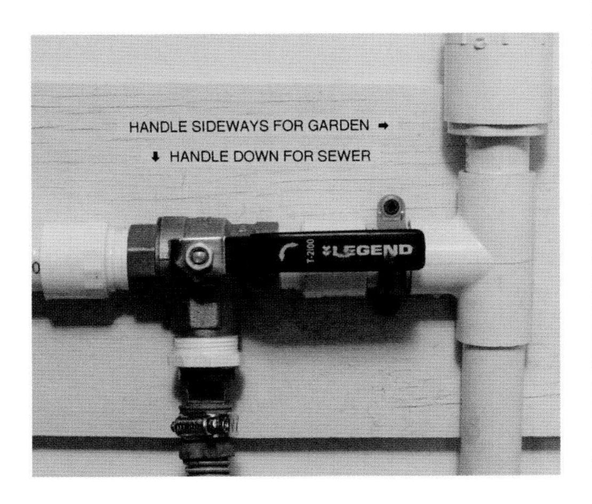

CHECKLIST FOR L2L INSTALLATIONS

Did you remember to:

→ Install the 3-way valve above the flood rim of the machine, in an accessible location, with a label?

→ Put the AAV on the greywater side of the 3-way valve and at the high point in the system, in a visible and accessible location?

→ Follow any necessary setbacks? (See Example Setbacks for Greywater Irrigation Area on page 57.)

→ Use 1" pipe and tubing with 1" × ½" tees to direct greywater to specific plants or mulch basins?

→ Irrigate appropriate plants based on how much greywater you produce and how much the plants need?

→ Leave an end of the 1" main-line tubing fully open, with no valve or cap?

→ Protect your washing machine by not overtaxing it with long irrigation runs? Remember to stay within the general safe distances for your washer (around 50 feet in a flat yard, not uphill).

→ Confirm that all greywater soaks quickly into the mulch? If there was pooling or runoff, enlarge mulch basins or add more outlets.

MAINTENANCE AND TROUBLESHOOTING

Laundry-to-landscape systems need maintenance about once a year, in addition to routine visual inspection of the plumbing parts and connections near the washing machine.

3-WAY VALVE AND AAV. Do a quick visual check for leaks frequently (and make sure value label is in place). Replace the AAV if it's leaking.

PIPING AND TUBING. If damaged, cut out the damaged section and replace with a coupling (for tubing use a 1-inch barbed coupling).

GREYWATER OUTLETS. Check for even distribution. Remove lint or hair built up in the outlets. Open ball valves and unclog with a small stick (there may be a glob of gunk in the valve). If needed, flush the system with a garden hose using one of the methods described on page 106.

If an entire section of the system is not receiving water, there may be a clog or kink in the main line; a clog could occur at a tee or 90-degree elbow. Dig up the section of tubing where the fitting is located, remove it, and check inside for debris. Or, examine the tubing anywhere it may have been kinked or damaged.

MULCH BASIN BELOW GREYWATER OUTLET. Check for signs of pooling. As mulch decomposes, the basin will drain more slowly. Remove and compost decomposed mulch and replace it with new mulch.

WASHING MACHINE NOT EVACUATING ALL THE WATER. Determine whether the problem is connected to the greywater system or if it also occurs when the washing is discharging to the sewer/septic system. Divert greywater to the sewer/septic system and observe the machine. If the problem still occurs, it's not related to the greywater irrigation system.

- If you suspect the 3-way valve itself may have clogged, disconnect the washer hose to unclog the valve.
- If the problem is not the valve and occurs on both the sewer/septic and greywater systems, the most likely cause is that the internal pump filter on the machine has clogged; clean the pump filter (see page 86).
- If the problem occurs only when the greywater system is turned on, check the outlets and end of the line for clogs. Also review your design and ensure you didn't overwork the pump by using too-small tubing (anything less than 1" for the main line) or by traveling too high uphill, or too far.

Irrigation Options

Both the L2L and branched drain system use mulch basins to soak greywater into the ground to irrigate plants. This section discusses how to construct a mulch basin for both types of systems, how to use a second 3-way valve in the landscape to create two zones, and (L2L only) how to irrigate raised garden beds.

How to Dig a Mulch Basin

Dig mulch basins in the "drip line" of the plant you'll irrigate, the area under where the branches end. Basins can be shaped like a circle, a semicircle, a trench, or a sun shape (see page 47), depending on where your plants are located. Plants growing along the fence or property line will have a semicircular or trench-shaped basin, whereas a tree in the middle of the yard may have a circle around it. Greywater spreads out in the bottom of the basin; larger basins wet more of the root zone.

Digging a trench-shaped mulch basin in front of a row of small citrus trees.

Note that the drip line of young plants will change as the plant grows. You can locate the basin at the current drip line or locate it a few feet away and supplementally irrigate the plant for the first year (until the roots reach the greywater).

If you encounter large roots, dig around or under them. Damaging small roots is like pruning small branches; it won't hurt the plant. Use the soil dug out of the basin to make a berm, creating a wall around the basin. The berm defines the area and reduces digging — increasing the height of the berm increases the depth of the basin. Stomp on the berm or use a tamper to compact the soil.

Alternatively, if you don't want a berm, put the soil in an out-of-the-way location or consider using it to elevate the surrounding area, since sunken basins are more water-efficient. Keep the bottom of the basin flat, or gently sloping downward to encourage water to spread through it.

Remember, multiple basins are better than one larger basin, and if you want to oversize the basin for an added safety margin, make sure to add more outlets to spread out the water.

In downward-sloping yards, dig a trenched-shaped mulch basin on the uphill side of the plant; use the excavated soil to create a berm on the downslope, to create a flat area uphill of the plant to infiltrate greywater. Be sure to tamp the berm well, and stabilize it as necessary, so rain won't wash it down the hill. See Design Considerations, page 81.

Add finishing touches to your basin: the sides of sunken basins can be stabilized with rock or other locally available materials (wood, sections of tree trunk, etc.).

Though L2L and branched drain systems both irrigate in a mulch basin, there are some specific differences:

L2L BASINS typically are 6" to 12" deep, with the deeper (12") basins for larger plants receiving more water.

BRANCHED DRAIN BASINS typically are 8" to 18" deep, based on the depth of the greywater pipe where it enters the basin. Dig 8" to 10" below the pipe so there is enough space for greywater to flow through an air space and land on a thick mulch layer. These basins are usually deeper than those for an L2L system because the branched drainpipe is usually buried deeper (due to the downward-sloping pipe). You may need to deepen the basin depending on where the pipe enters, since you dig the basin first and run the pipe to it.

Remember that the total area of the mulch basins must be sufficient to soak all your greywater into the ground; the size is determined by your soil type and the amount of greywater your home generates (see Sizing Mulch Basins on page 42 for details).

Three variations of mulch basins (filled with wood chips when complete)

How to Make a Mulch Shield

When greywater is distributed subsurface, roots can grow up the pipe and clog it. To prevent this, use a mulch shield to create an air space around the outlet; roots won't grow through air. Try to keep subsurface irrigation as close to the surface as possible — deeper is not better (see Subsurface or Surface Irrigation? on page 84).

The specific depth of distribution is determined by your state's code; for example, California requires each greywater outlet to be 2" below the top of a mulch shield. Use a manufactured irrigation valve box for a sturdy, long-lasting mulch shield. A free and eco-friendly alternative is to use repurposed 1-gallon plastic pots, though these will collapse if stepped on and won't last many years. Five-gallon buckets can also be used with branched drain systems or high-flow L2L systems. Another affordable and sturdy option is to use small sections of a wide-diameter (4" or larger) plastic drainpipe. Water-efficient washing machines can use smaller mulch

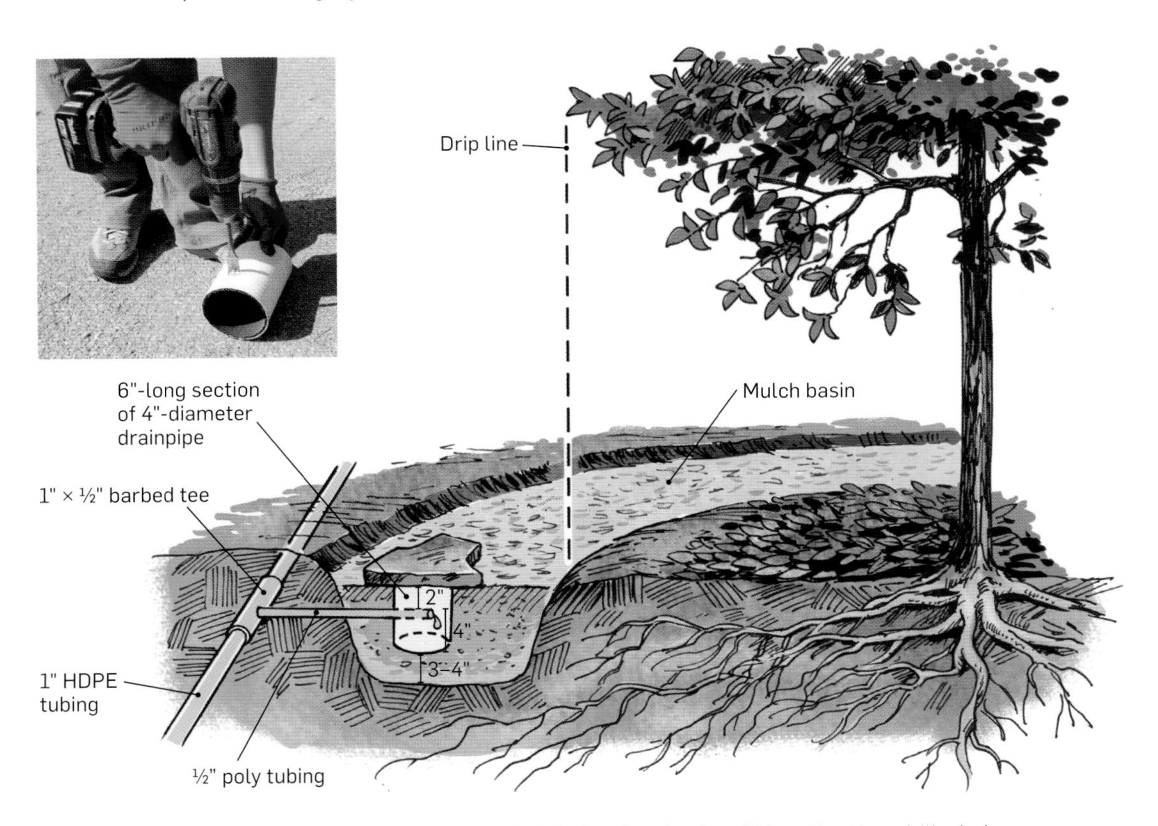

MAKING A MULCH SHIELD FROM 4" DRAINPIPE. *Cut the pipe into 6" lengths, then drill a hole 2" down (see photo) just large enough to insert the greywater tubing. Put something on top to cover it, like a flat rock. This can be used with water-efficient machines or pumped systems.*

MAKING A MULCH SHIELD WITH IRRIGATION VALVE BOX OR STURDY PLASTIC POT. *Position the pot or valve box so the wider end is down. Drill a hole large enough for the tube 2" below the top of the mulch shield. Cut off excess plastic below the hole — usually 4" to 6" below the hole is desirable.*

If using a sturdy pot or a plastic bucket: Cut an access hole, wide enough to fit your hand inside, in the bottom of the pot (positioned at the top when installed). Cut without compromising the strength of the pot. Place a flat rock or paver above the mulch shield to cover it.

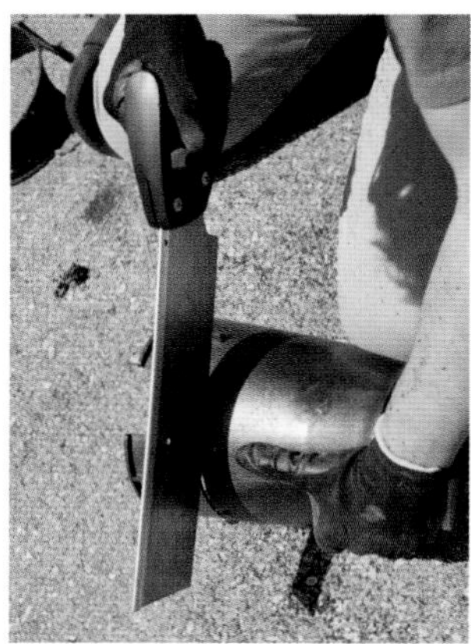

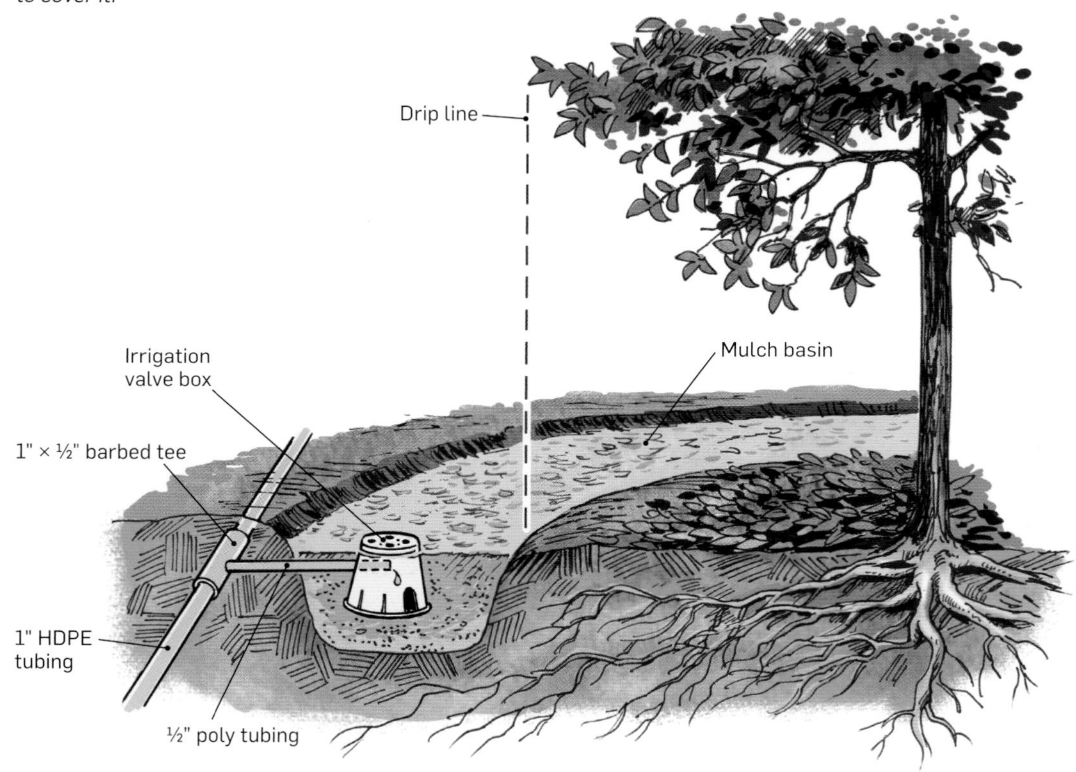

shields than top-loading machines, since each outlet will receive less water.

The important thing is to make sure greywater falls through *air* for several inches and onto several inches of *mulch*. Prepare the shield as described, then fill the mulch basin with a few inches of mulch. Place the mulch shield inside the basin, on top of mulch. Insert the greywater tube into the shield and fill the basin to the top with mulch. Maintain the air space between the outlet and mulch to prevent future clogs. You can add a flat paving stone or a garden statue on top of the mulch shield for aesthetics and to make it easier to find the outlets for annual maintenance.

Multiple Irrigation Zones

If your house produces lots of greywater and your plants are spread over different areas of the yard, consider two irrigation zones. Zones spread greywater over a larger area, and they require manual switching. If switching between zones on a regular basis for the next 10 years seems unrealistic, skip the zones and add more plants to benefit from the extra water (or divert excess to the sewer).

To add a second zone, install a 3-way valve where it's easy to access (for switching between zones). Use three 1" male (MPT) × barbed adapter fittings, wrapped with pipe thread tape and threaded into each side of the valve. Connect the greywater line from the machine to the middle port, then connect zone 1 and zone 2 to the right and left sides of the valve. It is possible, though a little more complicated, to install this second valve inside your house.

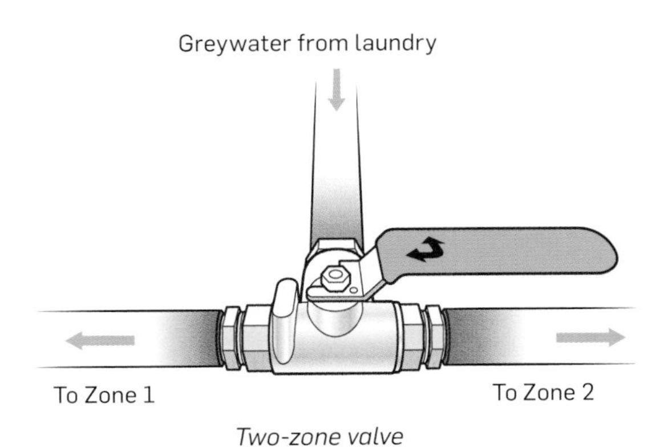

A second 3-way valve requires manual switching.

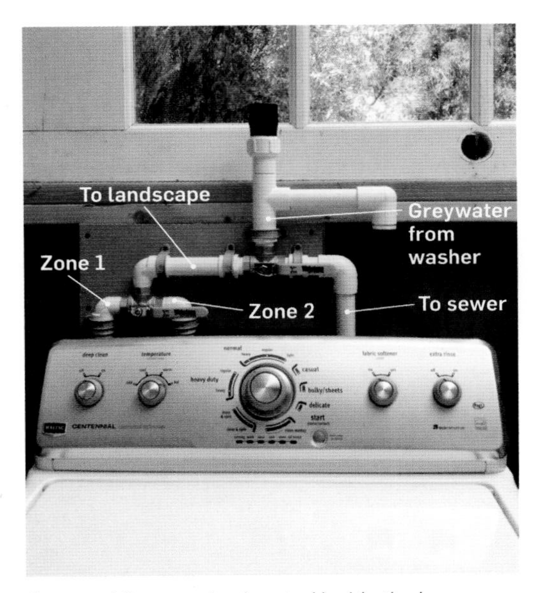

A second 3-way valve located inside the house creates two zones. The first valve controls the flow of greywater to the landscape or the sewer, and the second valve switches between two zones in the landscape. Note that two pipes will exit the laundry room.

Note: A similar technique can be used in a branched drain system using a larger 1½" or 2" valve.

You can also use an electronic actuator attached to a larger Pentair or Jandy type valve (see pages 142–147) with a battery timer to switch between zones on a regular basis. This method requires an outlet nearby for the actuator.

Irrigating Raised Beds

Since raised beds are harder to irrigate than in-ground plants using an L2L system, only include beds if other plants do not need irrigation. Take care to avoid kinks when directing tubing up into a bed: you may need to use two 1" barbed 90-degree elbows to safely enter the bed. If you plan to irrigate other plants in addition to the raised bed, send the 1" line into the bed and then back out, since water reaching a tee before the bed will take the lower route and not travel up the tube into the bed.

There are two common ways to irrigate raised beds. The first method is to use 1" × ½" tees, as described for other plants, directing water into mini-mulched areas in the bed. Plant large annual vegetables around the outlets (never root crops). The second method requires frequent maintenance (every few months) so is less desirable: Run the main line into the bed, cover the bed with straw, drill ¼" holes into the 1" tubing, and cover the whole thing with a half (semicircle) of a 3" drainpipe, as shown. This method allows greywater to be distributed over a larger area. However, the ¼" holes will clog and have to be cleaned out by hand — not something I want to spend my weekend doing.

MAINTENANCE-INTENSIVE OPTION. *Drill ¼" holes into the 1" tubing. Holes clog over time and require manual declogging. Cover tubing with drainpipe for code compliance.*

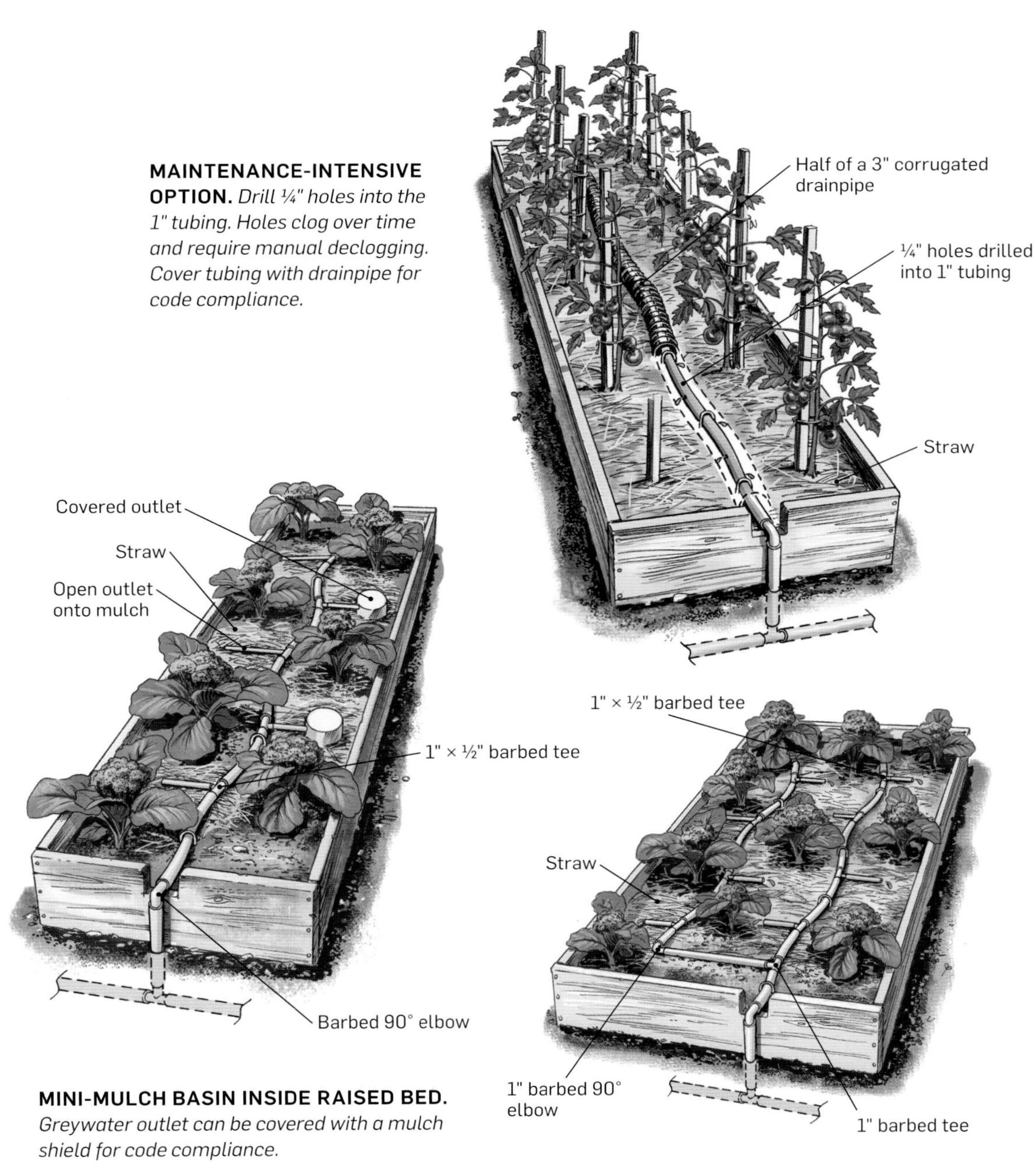

MINI-MULCH BASIN INSIDE RAISED BED. *Greywater outlet can be covered with a mulch shield for code compliance.*

Layout for a wider bed

CHAPTER 9

Install a Branched Drain Gravity-Flow System

A branched drain system (also developed by Art Ludwig) conveys greywater by gravity to the garden.

This system is commonly used with showers, sinks, or combined flows. A diverter valve taps into the drainage plumbing below a greywater fixture (e.g., a shower or sink). The diverter value can direct the water to the sewer/septic pipes or to the landscape. The user can control the flow of greywater by turning the valve's handle, either manually or remotely. Rigid, 1½-inch or 2-inch ABS pipe delivers greywater throughout the system (see Your Home's Drain, Waste, and Vent System on page 17 for more information).

Once in the landscape, the pipe divides repeatedly to distribute the water into multiple mulch basins to irrigate specific plants, including trees, shrubs, or large perennials. The cost and difficulty of installing branched drain systems vary greatly. Sometimes major plumbing work is required to access the greywater, while other times it is quite simple to divert the water.

In this chapter I'll lead you through the design and installation steps for installing your own branched drain system. We'll discuss installation details for the diverter valves, flow splitters, and (optional) electronic actuators.

IN THIS CHAPTER:

Design Considerations

The design of your branched drain system will be largely determined by your site factors, including where you tap into the plumbing, where the greywater pipe exits the building, and which plants you will irrigate. A careful assessment of your site, in combination with understanding how the system functions, will help you design a well-functioning system for your home. A successful design really pays off; once installed, these systems are extremely easy to use and maintain. The following section discusses each aspect of the system design to prepare you for construction.

If your home generates large volumes of greywater, make sure to read Got a Lot of Greywater? Don't Dig One Big Basin

Shower/tub water here irrigates six fruit trees. Fractions show how the flow splitters divide greywater flows into halves, quarters, and eighths.

(page 122) for advice on appropriately distributing larger flows.

Branched drain systems are well suited for kitchen sink greywater (note that not all states allow reuse of kitchen water). Be aware that kitchen greywater is more attractive to raccoons, bears, and other wildlife, and they may dig up your basins. Distribute kitchen greywater to multiple outlets to minimize maintenance; see pages 15 and 16 for more information.

Site Assessment

A branched drain system relies on gravity. This means that all of the irrigation piping must slope downward, and the landscape must be lower than the greywater drain pipes in the house. The minimum slope for the pipe is 2 percent, or ¼ inch of drop per 1 foot of distance traveled. This requirement can make it difficult to go under or around patios or walkways due to elevation loss. It's much easier to

GOT A LOT OF GREYWATER? DON'T DIG ONE BIG BASIN.

A common mistake for new greywater installers is to direct lots of greywater into one giant mulch basin. It may seem like digging one larger basin would be just as good as two smaller ones, but it's not. The reason? Though a large basin has lots of area, greywater flowing out one outlet will decompose mulch and slow the infiltration in the area surrounding the outlet, which can result in pooling greywater. To provide extra capacity in a system, it's better to divide the flow to more outlets, rather than to dig larger basins around fewer outlets.

install this system in downward-sloping yards than in flat yards. With the latter, you can opt to irrigate near the house, within 20 feet or so, but keep in mind that the pipe gets deeper and deeper the farther it travels.

Branched drain systems are best for irrigating trees, shrubs, and large perennials. If you have extra water, consider planting more to benefit from the greywater.

Ask yourself these questions to determine whether your site is suitable for a branched drain system:

1. Can you access the greywater source, install a diverter valve, and exit the house to reach the plants? Valves located in the crawl space can be turned by a motor (called an actuator) controlled with a switch in the house.
2. What plants will you irrigate with the system (e.g., trees, bushes, larger perennials)?
3. Are there any significant design challenges, like sidewalks, large tree roots, buried sewer, gas, or water lines?
4. If your system will be permitted, find out any special requirements that will impact your design.

Then, identify the plants to irrigate based on the quantity of greywater and the plants' water requirements (see How Much Water Do My Plants Want? on page 49).

Diverter Valve and Pipe Runs

There are many different plumbing configurations and potential challenges for installing the diverter valve. The basic steps are covered here, and you can adapt the concepts to fit your situation. If you are a plumbing novice, this is a great time to get help or hire someone to install the valve.

First, determine where to locate the valve and where you can exit the house with the greywater piping. This may limit your irrigation options. Also decide whether you need an actuator (the valve should be readily accessible, so if you'd need to slither in the crawl space to reach it, install an actuator so you can operate it remotely). After you've determined the valve location and where the greywater pipe will exit the house, confirm that it's still feasible to reach the plants you've planned to irrigate. If not, reassess the landscape.

Warning: Always divert greywater *before* a connection to a toilet drain.

If the drainpipes are in a crawl space and the system does not require a permit, install the diverter valve high enough so the greywater pipes can exit the house above the foundation. For permitted systems you must install the valve after the vent and P-trap. Some local authorities may allow you to deviate from this if you have a compelling reason, but most likely they won't.

In some situations it is more convenient to install the diverter in the tailpiece or trap arm (before the trap or vent), such as:

- When there is not enough elevation drop for the greywater pipe to exit above the foundation; for example, if the crawl space is very shallow. If the pipe won't be able to travel by gravity over the top of the foundation, you will need to core-drill through the foundation wall, which adds cost and time.
- When there is no room to fit the valve into the existing plumbing before it joins the toilet drain. It may be easy to install the valve before the trap, while installing it downstream may require extensive reconfiguring of pipes.
- In a sink-only system: the valve can be installed under the sink in the tailpiece or trap arm. This allows you to operate the valve from inside the house. Alternatively, open the wall and install the valve in the vertical drain just below the vent, and keep it accessible by creating an access hatch (greywater codes typically will permit this).

The standard diverter valve is about 6½ inches long and needs a little extra room for the transition couplings, so plan to remove about 7 inches of pipe to install the valve. However, it may be easier to remove a whole section of pipe than to cut into it, especially if the pipe is cast iron. If space is an issue, purchase a "space saver valve" (made by Jandy; see Resources); it's a few inches smaller than the standard model. If your system requires a backwater valve, you'll need several more inches of space.

Ideally, the greywater pipe exits the building above the foundation (for ease of installation). Remember, the pipe must drop at least ¼ inch per foot of horizontal run. Measure the distance in feet from the valve to where you will exit the house, and divide by 4. This represents the number of inches of drop required. For example, if you have 20 feet between the valve and the house exit point, the pipe run needs 5 inches of drop. Use a laser level or measure down from the floor joists to determine whether you have enough room for the elevation drop. If you can't exit above the foundation, reconsider the valve placement or plan to core-drill through the foundation wall (you can hire a company to do this, or rent equipment to do it yourself). And don't forget to check for foundation vents or crawlspace access doors to exit from.

PERMIT TROUBLES WITH 3-WAY VALVES

Sometimes inspectors who haven't permitted many greywater systems before may be hesitant to allow the typical Jandy or Pentair type 3-way valves because the valves are not made for drainage plumbing; they're designed for pool and spa applications. If your project requires a permit, the following info may help clarify the situation and your options with your inspector.

- The valves are certified (NSF/ANSI Standard 50; see Resources).
- An alternative to a 3-way valve is to use two 2-way ball valves, but this would allow the user to leave both sides open or both sides shut — clearly less desirable than using a valve made for pools.
- These valves work. They've been used all over North America for greywater with no apparent problems. Hundreds of these valves have been used in permitted greywater system projects with no issues.

As of 2015, there is a UPC-approved diverter valve for drainage plumbing called the GreenSmart Diverter. This valve is constructed from other plumbing parts but went through expensive testing to earn its listing. Many greywater installers are unhappy with the valve. Here's why.

Problems with the GreenSmart Diverter valve:

- Its high cost: $500
- Its large size makes it unsuitable for many retrofit installations. This valve requires about 16 inches of straight pipe, while Pentair and Jandy valves needs only 7 inches.
- Its lack of track record of functioning well in the field. Greywater Action has a valve that we use for demo in our trainings, and the fourth time the valve was used it spun endlessly. Had this occurred under the house no one would have known until the motor burned out.
- It's constructed with a ball valve inside, so when it's turned to greywater the ball valve shuts off the sewer pipe and greywater backs up before it flows out the greywater outlet, creating a place for crud to build up inside the valve just behind the ball. The manufacturer recommends turning the valve once a month to prevent this, but who will actually remember to do this?

As for the advantages to this valve over the Jandy and Pentair types, it's remote controlled and comes with two remotes so you can control it from two separate locations, and it's UPC-listed for drainage plumbing.

The company recently released a manual version of the valve ($75), which has the same disadvantages listed above, without the advantage of the remote control.

Splitting the Flow

Distributing water to the landscape with a branched drain system is different from other irrigation methods. A fitting called a **flow splitter** (also called a double-quarter bend, twin 90, or double-ell) divides water evenly in two directions. Use multiple flow splitters to divide the flow into halves, quarters, eighths, or sixteenths. For example, if 20 gallons of greywater from a shower flows through the system and is divided (via flow splitters) into eighths, each of the eight outlets discharges 2½ gallons.

A challenge with this system is that it's easier to send more water to the plants in the beginning of the system than the end. For example, a thirsty banana patch near the beginning of the system could get half the water from the first split. The image at right offers one method to send more water farther down the line of your system.

Make sketches of your system to plan how to divide the flow appropriately.

Note: Conventional wye or tee fittings also are able to divide the flow, but they don't necessarily do this evenly, so you wouldn't know how much water the plants are receiving. If you can't find flow splitters, use tees instead.

I recommend having one (or more) flow splitters on hand while you design the system; this helps with visualizing how to split up the flow appropriately. A common beginner's error is to use them like the tees used in a laundry-to-landscape system, inserting one into the line any time you want to irrigate. This won't work!

In order for the flow to split evenly, you'll need about 2 feet of straight pipe leading into the flow splitter. If the pipe curves just before the split, turbulence will send more water out one side than the other. In practice, it may not be possible to have exactly 2 feet, but do the best you can. Very steep slopes above the splitter may also alter the evenness of the split.

In most systems, a single shower should have at least four to eight outlets, a kitchen sink four to six, and a bathroom sink two to

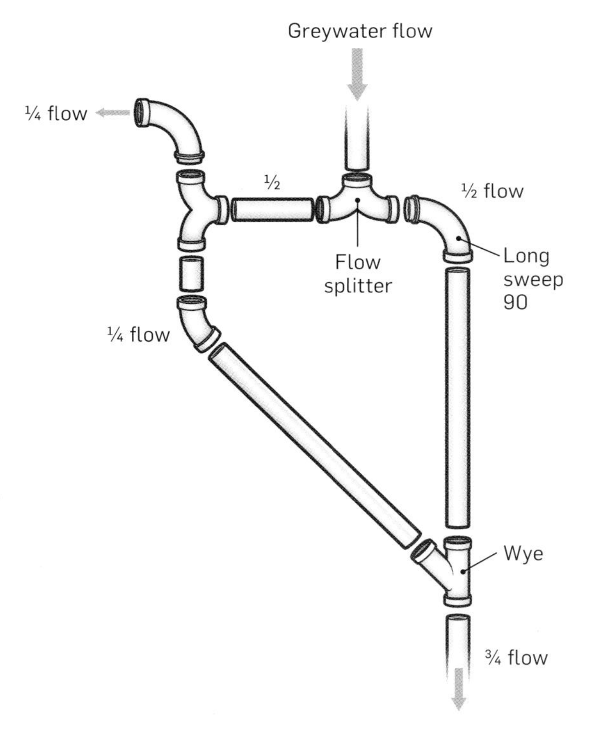

Loop-de-loop pipe configuration directs more flow farther down the system, where it can be divided more as needed.

SITE PLAN OF DIVIDED FLOWS

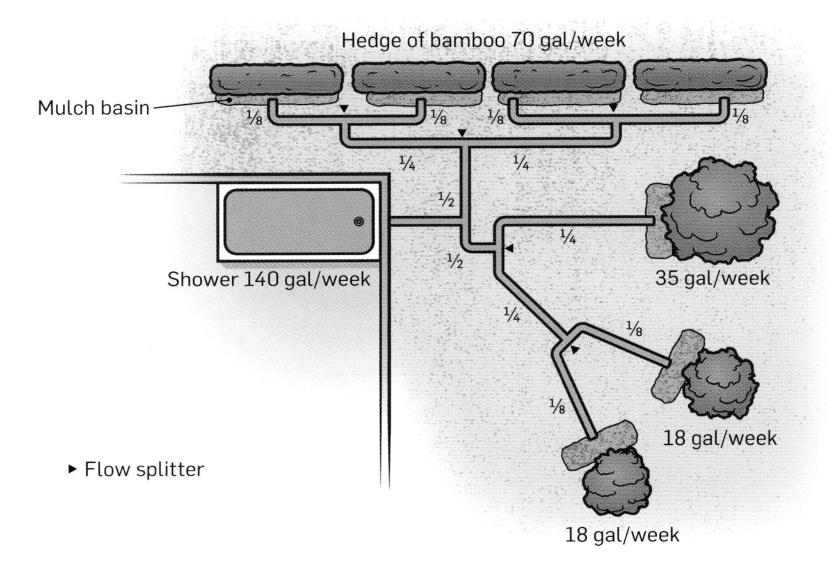

three, depending on usage (more outlets for more people/more greywater in the system).

Draw a sketch of your system as accurately as possible. After you install the system, return to this site plan and adjust it as needed so the "as-built" site plan accurately shows where the pipes and outlets are. Put this in your O&M manual (see page 64), along with photographs of the unburied system.

Backwater Valve

Permitted systems likely will need a backwater valve, a one-way check valve to prevent sewage from backing up into the greywater system. In practice, this is a redundant safety precaution; the 3-way valve physically blocks off the sewer side when it is turned to greywater. However, if the valve isn't shut properly, backed-up sewage could enter the system. If you install a backwater valve, keep in mind that a plumbing snake (used to unclog a sewer line) can't come back through the valve without breaking it; be sure to show the valve to any plumber working on the pipes in the future.

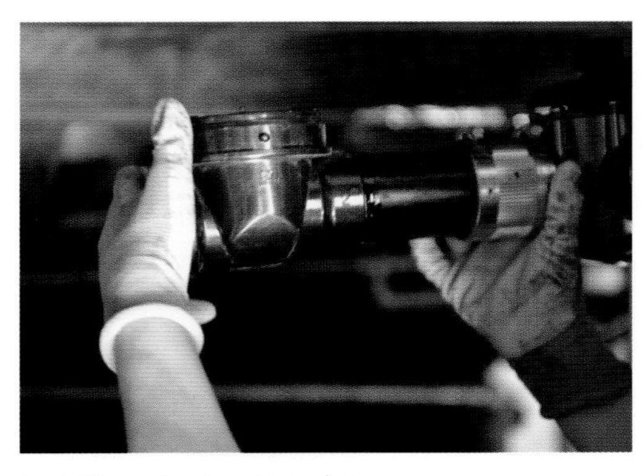

Installing a backwater valve

Installing a
BRANCHED DRAIN SYSTEM

This project walks you through the basic steps of installing a branched drain system. No two branched drain systems are identical, so you'll likely need to adapt the steps to fit with your home and landscape. Installing a branched drain system typically takes two or three times longer than installing a laundry-to-landscape system. Cutting into old plumbing can lead to unexpected complications, working in a crawl space is slow, and sloping the pipe is very time-consuming (plus a lot of digging!). Having friends or hiring someone to help will make the installation easier.

The bulk of the materials you'll need are available at any plumbing store, but the diverter valve and flow splitters most likely are not. Order those parts a few weeks before you plan to install the system. If you'll be getting free wood chips from a local tree trimmer, allow a few weeks for the delivery. See Plumbing Basics for Greywater Installation (page 178) for details on plumbing parts and essential techniques.

Materials

- One 1½" × 2" 3-way diverter valve (Pentair or Jandy brand)
- One 1½" or 2" ABS backwater valve (as needed)
- Transition couplings (to install diverter valve to your existing plumbing)
- 1½" or 2" ABS pipe and fittings (as needed)
- ABS straps (to hang pipe in crawl space), or 2-hole straps or plumbers tape (to strap pipe to walls)
- Cleanout plugs (as needed)
- Flow splitters (twin 90s or double-quarter bends; see step 6)
- Actuator and related parts (for automatic switching via a switch inside the home; optional)
- ABS glue (solvent cement)
- Irrigation valve boxes (as needed; see step 10)
- Construction-grade sealant (see step 11)

Tools

- Pipe cutter or hacksaw
- Cutter compatible with your existing plumbing (see Pipe Materials for Drainage Plumbing on page 181)
- Hex-head (nut) driver
- Tape measure
- Permanent marker
- 2-foot level and torpedo level (with marks for 2% grade)
- Tin snips or drill, piloting bit, and hole saw (see step 4)
- Digging tools (mattock, trenching shovel, digging bar, etc.)
- Painting supplies and caulking gun

1. INSTALL THE 3-WAY DIVERTER VALVE.

Depending on your plumbing setup, you'll probably want to alter the 3-way diverter valve to change the location of the inlet. The valves are preset to use the middle port as the inlet, but usually it's more convenient for the inlet to be on one of the side ports. Take the valve to the drainpipe and determine whether you should change the inlet. See 3-Way Valves for Branched Drain Systems (page 130) for installation instructions.

Measure and cut out a section of the drainpipe to insert the diverter valve. If you're using a backwater valve, consider where it will go before cutting the pipe (see Two Valve Orientation with Backwater Valves on page 129). Most backwater valves must be installed in the horizontal position.

Note: If your house has old metal plumbing — prone to leaks when jiggled — make sure you feel confident to handle other plumbing problems that could arise from installing the valve.

Connect the valve with no-hub or transition couplings: Loosen the metal band and slide it over the pipe. Then, fit the rubber coupling over the pipe and valve, bending it back for a tight fit. Slide the band over the coupling and tighten the band with a hex-head driver.

Step 1. Hold the 3-way valve in position next to your greywater pipe below house.

TWO VALVE ORIENTATIONS WITH BACKWATER VALVES

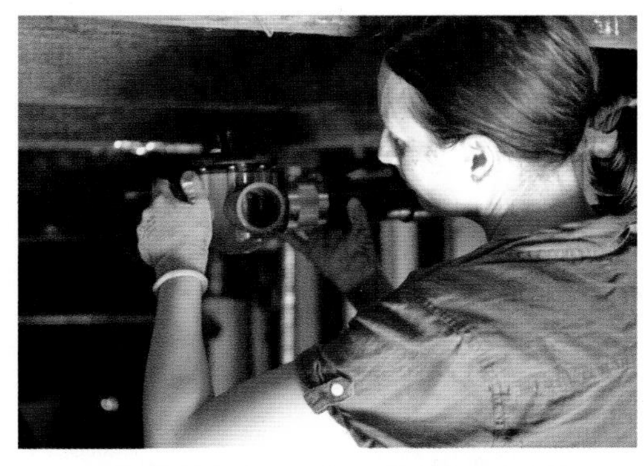

Installing the 3-way valve. Note that this valve was adjusted to use a side port for the inlet.

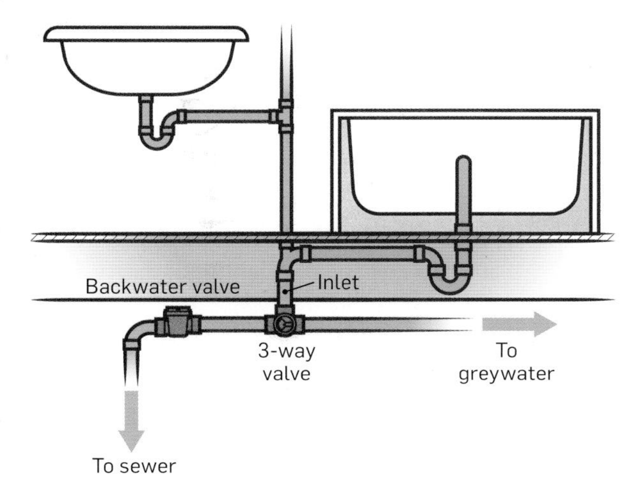

Example of diverter valve using the preset middle port for the inlet

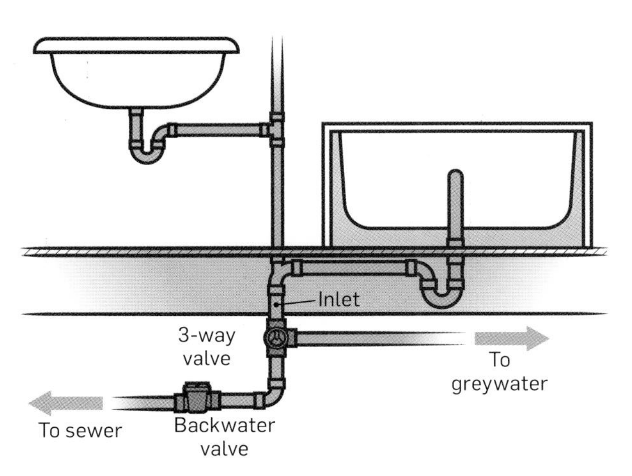

Example of diverter valve adjusted to use a side port for the inlet

3-WAY VALVES FOR BRANCHED DRAIN SYSTEMS

Plastic 3-way valves are made with the inlet in the middle port and an outlet port on each side. Unlike the brass 3-way valves used in an L2L system, these larger plastic valves can be altered so the inlet enters any port (see instructions, below). These valves are made for the pool and spa industry and often have to be mail-ordered. Two common brands are Pentair and Jandy. They are "never-lube" valves, meaning they're self-lubricating. Each valve works with two sizes of pipe: the 1½"–2" size works for both 1½" and 2" pipe. Jandy also makes a 3" valve and a "space saver" valve for tight spots.

Whenever possible, use transition couplings instead of glue to connect pipes to the valve. This protects the valve from being made unusable if you apply glue in the wrong place. If a valve location is hard to reach, install an actuator to turn the valve remotely. However, all valves, even with an actuator installed, must be accessible for future repairs (see How to Wire an Actuator on page 142).

How to Change the Inlet of a 3-Way Valve

Pentair and Jandy valves require different modification techniques.

PENTAIR VALVE. This is the easier of the two valves to change.

1. Remove the plastic "pins" on the handle, using a flathead screwdriver.

MODIFYING A PENTAIR VALVE

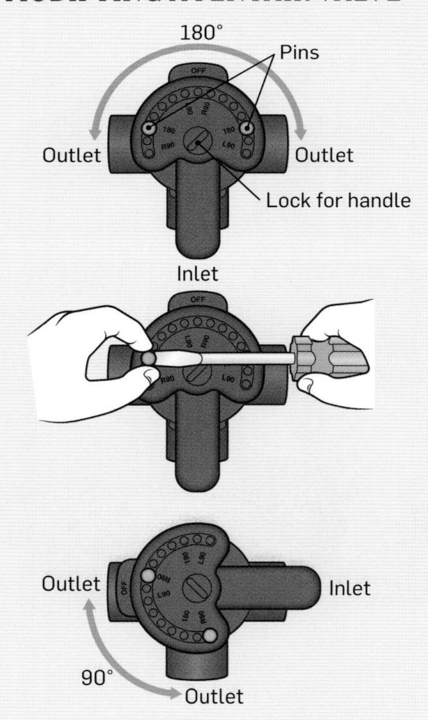

Handle swings in an arch, moving the "off" between ports.

2. Position the valve so the handle shuts off an outlet port of your choice. The valve turns 180 degrees when the inlet is in the middle, or 90 degrees with the inlet on a side. Place the pins in the desired location (right 90, left 90, or 180).

3. Test the valve by turning the handle and observing which ports it shuts off. If it doesn't turn as desired, switch the pins (e.g., from the left 90 to the right 90 position).

JANDY VALVE. Be careful when adjusting this valve. If the handle and internal shutoff become misaligned, it can become tedious to fix.

1. Remove all screws on the top of the valve. Using a flathead screwdriver, break the seal and lift the faceplate off the valve.

2. Carefully position the valve so the "inlet" label on the faceplate faces your desired inlet and the handle "off" position is shutting off either the greywater or sewer/septic pipe.

3. Reinstall just a few screws around the circle, and test the valve. If it works as desired, add the remaining screws.

Note: After a Jandy valve has been altered, the valve handle will turn past the shutoff outlet. Label the valve clearly to prevent moving the handle too far.

MODIFYING A JANDY VALVE

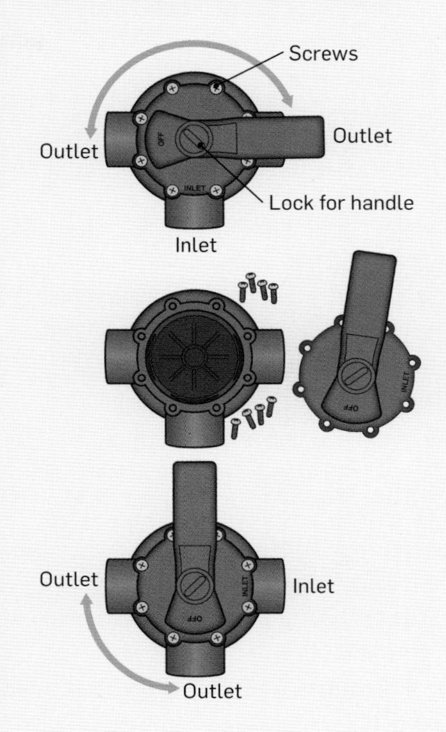

2. INSTALL AN ACTUATOR (OPTIONAL).

Install an electronic actuator if the valve is not easily accessible for manual operation. See How to Wire an Actuator (page 142) for installation instructions.

3. PLUMB TO THE EXIT POINT.

Complete the piping run from the diverter to the house exit point, using ABS pipe and following standard plumbing techniques and your local code requirements. Support the pipe every 4 feet, either with strapping fastened to something solid, or use rigid ABS pipe hangers attached to the floor joists to prevent vertical and horizontal movement of the pipe. Maintain a minimum slope of ¼" drop per 1 foot of run, checking the slope with a grade level.

Install cleanouts (see page 137), required every 100 feet of horizontal pipe, or after turning an aggregate of 135 degrees, such as with a 90 bend and a 45 bend. Use *long-sweep* fittings for horizontal-to-horizontal or vertical-to-horizontal turns to prevent clogs. Use *short-sweep* fittings only for horizontal-to-vertical turns.

4. RUN THE PIPE OUTSIDE.

When possible, exit through a crawl space vent. Cut a hole out of a corner of the vent screen that's large enough for the pipe, using tin snips. Or, drill a hole through the wall just large enough for the pipe (see page 93).

If the pipe is too low to exit above the foundation, you have two options:

- Bury a surge tank under the house and pump the water over the foundation. This is a good option if you have enough space for

Step 3: Tighten clamp on transition coupling (diverter valve shown with actuator installed).

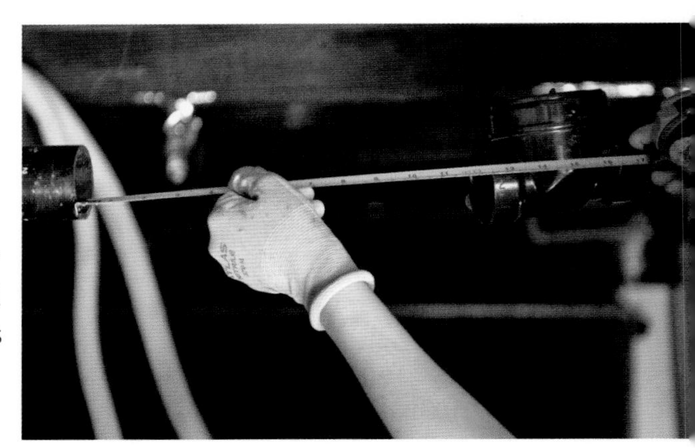

Measure to cut a length of pipe.

Short-sweep 90 (left) vs. long-sweep 90 (right)

the tank and you need a pump for the rest of your system (if your landscape is uphill or you'll be filtering the water for a drip irrigation system). If the rest of the system can be gravity-fed, go with the second option.

- Core-drill through the foundation. Hire a contractor or do it yourself with a rented drill and core bit. This will add some cost and labor but allows for a gravity system instead of a pump.

5. DIG THE MULCH BASINS.

See How to Dig a Mulch Basin on page 113.

6. MOCK UP THE OUTDOOR PIPE RUNS AND DIG THE TRENCHES.

On top of the ground, lay out the pipe and fittings to each basin. Cut pipe as needed to determine where the pipe and fittings will be located (don't dry-fit; just lay out the materials so you'll know where to dig the trenches). Add cleanouts to the flow splitters: Purchase flow splitters with cleanouts pre-installed, or you can make your own. See Installing a Cleanout in a Flow Splitter (page 137) for instructions.

Remove the first section of pipe so you can dig its trench. Use a mattock and trenching shovel to dig to the first flow splitter. In flat yards, bury the pipe as shallow as possible at the beginning; it must slope downward continuously and will get deeper as you go. If the yard slopes downhill, you can bury the pipe deeper at the beginning because it can stay at a relatively constant depth.

If cars will drive over the pipe, make sure there is a foot of solid soil above the top of

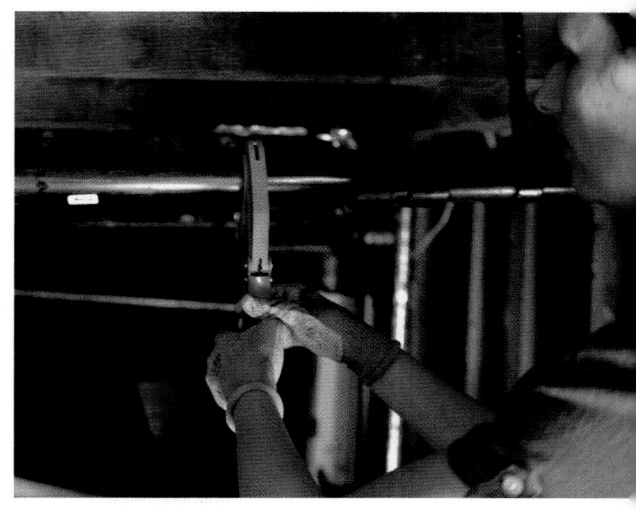

Cut ABS pipe.

Irrigation piping for branched drain system

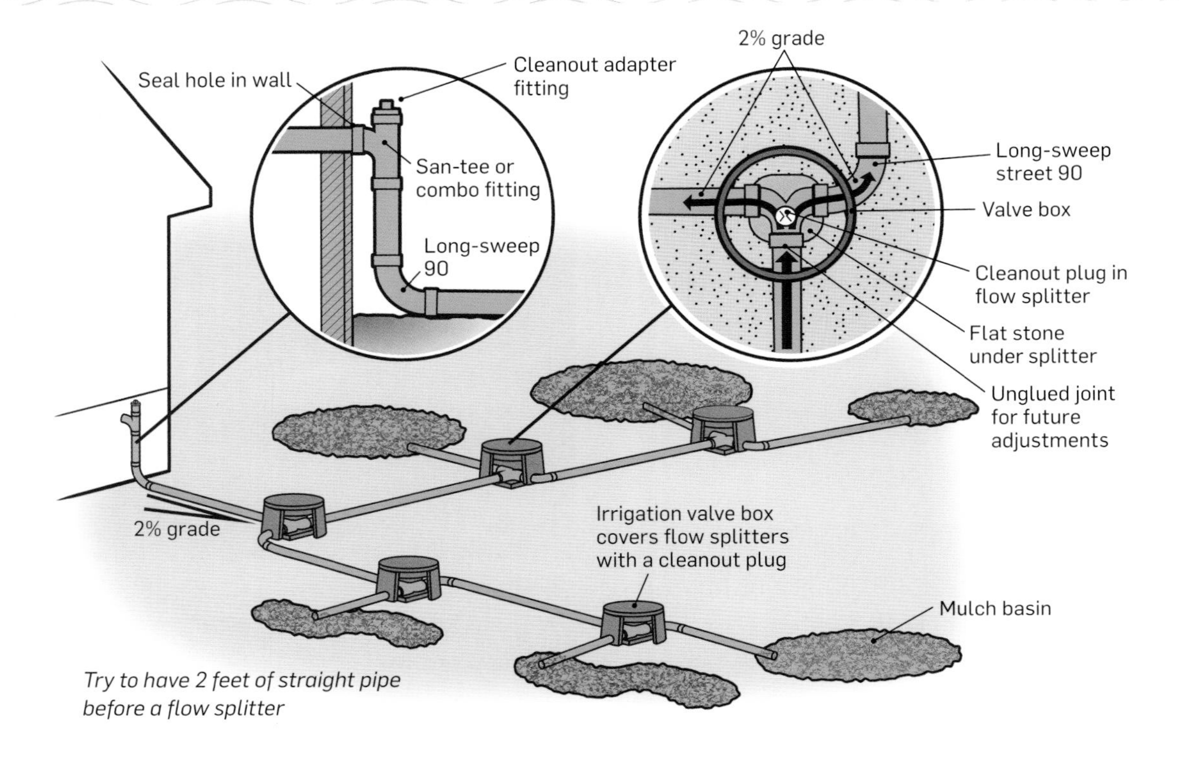

Try to have 2 feet of straight pipe before a flow splitter

Use a level to slope pipe at 2% grade.

the pipe. If you're getting a permit, talk to the inspector about burial depths; some may want to see the pipe buried deeply at the beginning, as they would for ABS carrying sewage, though these standards shouldn't apply to an irrigation system. If the inspector disagrees, you must argue your case or risk not being able to install this system; it won't work if the pipe has to be 2 feet underground.

Tip: When trenching through a lawn, use a flat shovel and cut out pieces of turf that you can later replace, then dig the trench with a mattock. The end result will be cleaner than a messy trench through the lawn.

Level the trench so the bottom slopes ¼" per foot and is as smooth as possible. It's okay to slope more steeply, but in a flat yard the accumulated increase in pipe depth creates a *lot* more work. Check for slope using a grade level. In very rocky soils you may need to bring in extra dirt to properly slope and compact the trench.

7. DRY-FIT THE PIPE RUNS.

Assemble the pipe and fittings needed to reach the first flow splitter. Use long-sweep 90-degree bends (not short-sweep bends) or 60-, 45-, or 22-degree bends to change direction in the pipe runs. To turn directly from a flow splitter, use a *street* fitting, which inserts into the hub of the flow splitter. If any pipe is warped, install it so the warp runs sideways (not up or down).

Position the flow splitter on a brick or flat stone to prevent it from tilting as the soil settles. The splitter must be perfectly level across the top; check it with a torpedo level. Slope the pipe out of the splitters: gently push the pipe downward with a level on it to get the proper slope.

Trench and dry-fit the next section of pipe. Continue trenching a section at a time until you've completed the whole system. You may need to weight down the middle of long, straight runs of pipe with soil or a large rock. Remember to maintain a ¼" slope the entire way. If you encounter roots or underground pipes, go under them: do not dip down and up again; otherwise, solids in the pipe will settle and create clogs. Check the depth of each mulch basin, and deepen any basin with less than 8" between the pipe and the bottom of the basin.

Step 7. Dry-fit the pipe.

REDUCING FROM A FLOW SPLITTER

If you started your system in 2" pipe you'll use a 2" flow splitter for the first split, but can reduce down to 1½" for the rest of the system. Here are some options to reduce out of the flow splitter:

→ The easiest way is to use a standard bell reducer (reducing coupling 2"–1½"). A bell reducer creates less of a clogging potential than a reducing bushing (which is another option) though both of these parts are made for use with greywater and don't have clogging issues under typical use.

Bell reducer 2–1½"

→ The best way is to find a 2" by 1½" by 1½" flow splitter (double quarter bend) — they are uncommon, but they do exist.

8. TEST THE SYSTEM AND GLUE THE PIPE RUNS.

Support the pipe by piling mounds of soil or large rocks over the pipe at intervals to hold it in place. Run water through the system and check the outlets for appropriately even flows in accordance with your design (outlets designed to receive ¼ flow should receive more water than outlets with ⅛ of the flow). Either visually observe the flows, or place containers (a mason jar or a 32-ounce yoghurt container size) under each outlet to measure the flow.

Adjust the angle of flow splitters (tilt them as needed to adjust the flow) or the pipe until water flows evenly from each pipe. So long as you have an even split of water in the flow splitter, the two pipe runs out of the flow splitter are independent of each other. Each pipe's slope and length won't affect the other pipe; for example, if one side of your yard slopes downhill, that side won't get more water than the other.

Optional: You can reinforce the system with a few strategically placed pieces of rebar and some wire, for example, where the pipes exit the flow splitters. This will help prevent settling of the pipe, which can throw the system out of balance. Here's how: Drive two short pieces of rebar into the soil on either side of the pipe, wrap a wire firmly around the rebar and the pipe, then hammer the rebar down until you've reached the desired slope. As you hammer the rebar, the wire will push down the pipe. Wait until you've glued the system before adding rebar and wire.

Mark the pipe and fitting at each joint so you glue them in the right position. Mark which side of the pipe should be on top — any warps should go left/right, not up.

Note: Don't glue the joint where the pipe enters into the flow splitter. If you leave this unglued you can more easily adjust the system if it gets out of level in the future.

Note: If the pipe sits in the sun, it can warp and change the slope on the system; be aware of this if you don't bury the pipe right away.

Keep flow splitters level.

Optional: Rebar and wire hold the pipe to maintain proper slope, even if the soil settles over time as it exits a flow splitter.

INSTALLING A CLEANOUT IN A FLOW SPLITTER

Flow splitters are manufactured as a solid fitting but can be easily adapted to include a cleanout plug in the center of the fitting. Just drill a hole into the fitting to add a plug. A cleanout is a handy addition for two reasons: you can remove the plug and observe the water flowing while testing the system, and you can check for clogs or insert a hose to flush the system. Because this alteration isn't watertight (which doesn't matter because water is flowing on the bottom of the fitting and the plug is on the top) and it "mixes materials" — a PVC plug into an ABS fitting — inspectors generally don't approve of it; ask if your system will be inspected. Alternatively, you could install a sanitary tee fitting with a cleanout cap immediately before the flow splitter, and use a larger valve box to cover both the cleanout and the splitter. (See image on page 138.)

To install a cleanout into the flow splitter:

1. Drill a 1¼" hole in the center of the flow splitter, using a drill and 1¼" hole saw (you have to secure the fitting before drilling so it won't spin; insert a piece of unglued pipe into the fitting and step on it to hold it down).

2. Cut threads into the hole by inserting a 1" metal nipple (a short piece of threaded steel pipe) and turning clockwise with tongue-and-groove pliers. Remove the nipple by turning it back out.

3. Screw in a 1" plastic plug a few turns (it doesn't need to screw all the way in).

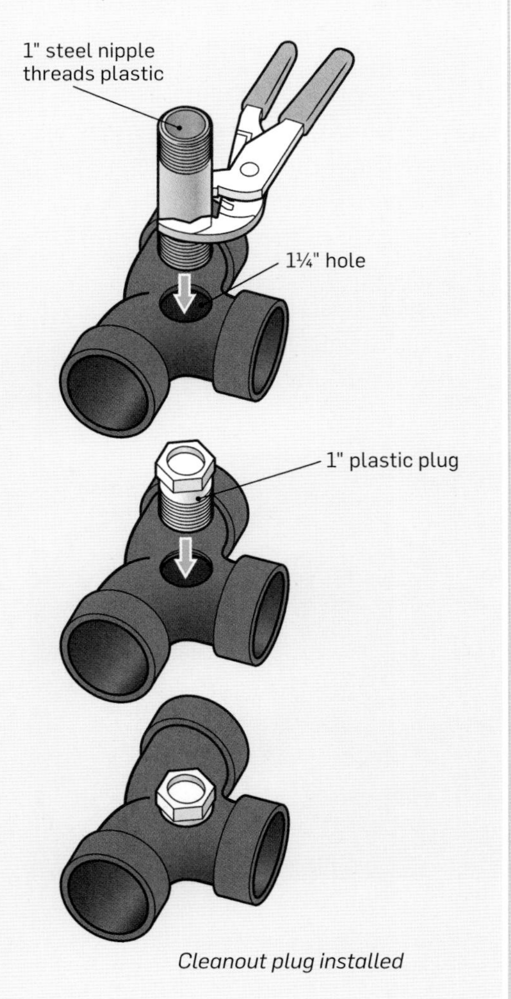

Cleanout plug installed

After the pipe has been glued, be sure to mound dirt on top every few feet to hold it down.

Go back and glue everything from the top to the bottom, except for the joint into each flow splitter. Keep the levels handy while gluing, making sure you don't over- or under-slope any sections. Conduct one more flow test, as needed, before burying the pipe.

9. ADD MULCH SHIELDS OR REINFORCE THE OUTLETS AT THE BASINS.

If the pipe enters the mulch basin below the mulch, add a mulch shield onto the end of the pipe to prevent it from clogging. The shield and pipe end extend into the center of the basin. See Making a Mulch Shield for a Branched Drain System (page 139) for instructions.

If the pipe enters the basin above the level of the mulch, reinforce the ground where the pipe emerges, using stones. The pipe should extend a few inches into the air to provide an air gap over the mulch.

An alternate cleanout method that doesn't alter the flow splitter. Use a sanitary tee on its back immediately before a flow splitter with a cleanout cap.

10. TAKE PICTURES AND BURY THE PIPE.

Photograph the entire system while it's exposed. If the system is permitted, inspectors will want to see it unburied. Often, they'll let you bury long straight runs of pipe with the fittings exposed, but be sure to find out; if you can't bury it, put large rocks or mounds of dirt in just enough places so the pipe won't move around.

As an optional step, you can wrap the pipe with wire so it's findable with a metal detector. Also consider using valve boxes with metal in the lid (most come this way).

Begin burying the pipe by packing fine, rock-free soil under and around the pipe, either with your hand or foot. If you bury it with loose dirt below, the pipe can settle and change the levels. Make sure you don't push up the pipe as you are packing the dirt. Double-check the levels before covering the pipe, making sure the top of the pipe is clean so dirt doesn't interfere with the level. Now bury the pipe completely while leaving the flow splitters exposed.

If you've included cleanouts in the flow splitters, make an access box for each. An irrigation valve box is easy to adapt for this: just cut notches in the bottom so it fits over the pipes and the flow splitter. Keep the lid of the valve box at the surface of the ground. If you haven't included cleanouts, it's still a good idea to use an access box. It's okay to bury them so long as you mark the location in case you decide to alter the system in the future or you need to adjust it if the splitter gets out of balance.

MAKING A MULCH SHIELD FOR A BRANCHED DRAIN SYSTEM

Make a mulch shield from an irrigation valve box or a 5-gallon bucket. Maintain a drop of several inches between the pipe and the mulch to prevent roots from growing into the pipe.

To create a mulch shield:

1. Put a layer of mulch under the outlet, about 4" to 6" thick.

2. Set the mulch shield on the mulch, and mark where the pipe will enter.

3. Drill a hole, slightly larger than the pipe, into the mulch shield with a hole saw. Carefully insert the pipe into the shield. Check the levels. Adjust the height of the mulch shield as needed; for example, if it raised up the pipe, wiggle the mulch shield down.

4. Mulch shields made from an upside-down plastic bucket need a hole cut in the bottom (which is now the top of your mulch shield) so you can see inside. Cut a hole large enough to see inside without compromising the strength of the bucket. Cover the hole with a flat paving stone.

After all mulch shields are in place, test the system again. If everything flows evenly, fill in the rest of the basin with mulch.

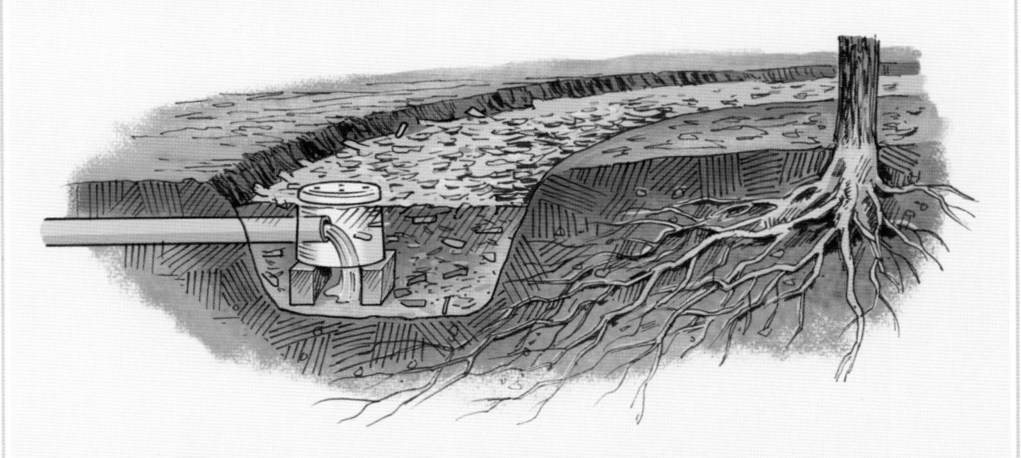

A mulch shield prevents roots from clogging subsurface outlets.

MAINTENANCE AND TROUBLESHOOTING

A properly installed branched drain system requires very little maintenance. Each year you'll need to check the system for proper flow and replace mulch as needed.

MULCH BASINS. Check the mulch basin under each outlet annually. If there is pooling greywater, remove the decomposed mulch under the outlet and replace with fresh mulch. If pooling occurs frequently, there may be too much water entering the basin. Enlarge the basin, add more plants to take up the water, or redesign the system so that less water enters that basin. After a few years, you may need to dig up more of the basin (not just under the outlet) and replace the mulch.

OUTLETS. Check outlets for flow annually. If the flow is uneven, first check the flow splitters for clogs. Flush the system with a garden hose inserted into a cleanout, or blast water backwards from an outlet. If the flows are still uneven, check the levels of the flow splitters using a torpedo level (they should be flat). To adjust the slope, turn the fitting until the level reads flat. If the flows are still uneven check the slope of the pipes leaving the flow splitters and readjust as needed. Note that if you glued the joint into the flow splitter you'll need to cut the pipe and add an unglued coupling to make the adjustment. If the pipes are under-sloped, solids may settle and create clogs, especially if the greywater is from a kitchen sink.

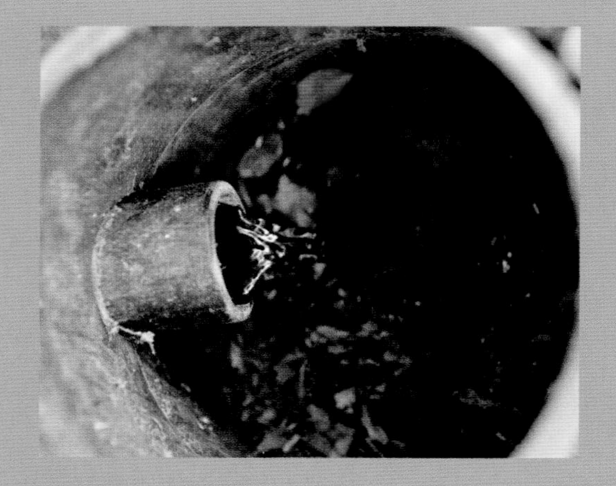

RACCOONS DIGGING UP BASINS. Raccoons are attracted to earthworms below the greywater outlets. If they dig up the basin, shovel the mulch back inside. If they disturb mulch shields, drive a long stake into the ground next to the outlet shield and screw the shield to the stake.

DIVERTER VALVE WON'T TURN FULLY. Unscrew the faceplate of the valve and remove it. Clear out any debris. Replace the faceplate, making sure it's in the correct position so the valve shuts off the greywater or sewer/septic pipes as designed.

11. CHECK FOR LEAKS, PAINT THE PIPE, AND ADD THE LABELS.

Run water through the system and check for leaks. Pay special attention to joints under the house in places you won't frequently see. Turn the diverter valve and check for leaks on the sewer/septic side.

Paint any exposed plastic pipe with exterior latex paint (the same color as your house) to protect it from sun damage. Add a label next to the diverter valve, and (if applicable) at the switch for the actuator. Seal any holes in the wall or floor with a construction-grade exterior sealant that is waterproof and adhesive to protect your home from water damage. Sikaflex is a common brand. Regular silicone caulk can peel away over time.

Note: For systems requiring a permit, inspectors typically want to see a "running test," with water flowing through the system. After the inspection, bury any exposed pipe.

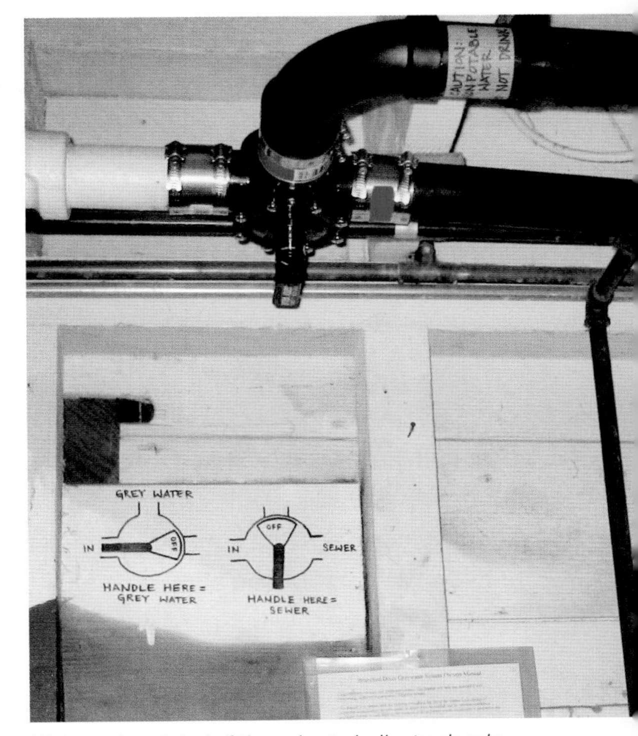

Make a clear label of the valve to indicate clearly where the water is going (to sewer; to garden), or take a photo of the valve with the handle in both positions and label the photo. Post the label prominently.

How to

WIRE AN ACTUATOR

An actuator, essentially an electric motor, attaches to the face of the 3-way valve and allows you to control the valve by a switch located inside the house. The 24-volt actuator is wired to a plug-in transformer, which provides power, and to a toggle switch to control the valve. Plan to locate the switch near any electrical outlet where you can plug in the transformer. Generally, the best location for the switch is near the fixture it controls, mounted on the wall of the bathroom or inside the vanity. Or, install it near any electrical receptacle with easy access to the switch; for example, at the entrance to the basement or inside a utility closet.

An actuator is needed on any diverter valve that is not readily accessible, such as in a crawl space, cluttered basement, or other out-of-reach location. The actuator must be accessible for repairs and not be buried inside a wall or under a floor (install an access hatch in these situations). Installing an actuator typically adds a few hundred dollars in materials as well as a few hours of labor. If there is no available receptacle to plug in the transformer, you'll need to have one installed, following local code requirements (in bathrooms and utility areas you'll likely need a GFCI-protected receptacle.) Anyone with basic wiring skills can install an actuator, but professional assistance may be needed if your site requires a new electrical outlet.

If you're inexperienced with actuators or your situation requires that you adjust the actuator (see step 1), it's a good idea to set it up inside the house and make sure it works properly before you attach it to the 3-way valve in the crawl space (either do this before you cut in the 3-way valve or use a second valve to practice on). It's possible for an actuator to be defective, and it's a lot easier to discover this if you're working with everything inside the house.

Materials

- One actuator that is compatible with your 3-way valve (common brands include Goldline and Pentair)
- One toggle switch (on-on, or on-off-on) (single-pole double-throw)
- One plastic electrical outlet box with solid cover
- One ¾" × ½" reducing bushing
- One waterproof strain relief cord connector
- 18 AWG 3-wire cable, 20 feet (or as needed)
- One 24-volt plug-in transformer
- Wire connectors (wire nuts)
- Wire clips

Tools

- Screwdrivers
- Wire strippers
- Needlenose pliers
- Drill with driver and bits

1. DETERMINE HOW THE ACTUATOR SHOULD TURN THE VALVE.

Just as the 3-way valve was manufactured to make a 180-degree rotation (with the inlet being the middle port on the valve), the actuator does the same unless you change it. If you changed the orientation of your 3-way valve so that the inlet is on the left or right side (see 3-Way Valves for Branched Drain Systems on page 130), you need to change the actuator. You can read the manufacturer's instructions on how to do this, though it's often easier just to play with it. Typically, you'll open up the actuator and readjust the cams, which determine how far the motor turns the valve in each direction. Since you can't test the actuator until it's mounted and wired, you may choose to wait until Step 6, Test the actuator, to make these adjustments, or adjust it on a second 3-way valve.

Step 1: Open the actuator to access the cams.

To adjust the cams, lift one up and reposition it to stop the motor in a 90-degree turn. Depending on the brand of your actuator, you may need to loosen a locknut to lift the cam. One cam stops the motor as it turns clockwise, while the other cam stops the counterclockwise rotation.

2. MOUNT THE ACTUATOR ONTO THE DIVERTER VALVE.

Attach the actuator to the face of the 3-way valve, following the manufacturer's instructions. Typically, you take off the valve handle, unscrew four screws on the faceplate (these screws line up with the screw holes on the actuator), then screw the actuator onto the valve with longer screws, and reattach the handle. Make sure the actuator's switch is turned on. (Turning it on at this step will save you a trip into the crawl space later on.)

Remove handle of 3-way valve.

Remove screws to mount actuator.

Screws removed to mount actuator

Mount actuator onto valve.

Actuator mounted onto valve with longer screws holding it in place

Fit valve handle onto actuator.

3. RUN WIRES FROM THE ACTUATOR TO THE ELECTRICAL OUTLET.

Run the actuator's wires to the switch location. For example, if you plan to locate the switch inside the bathroom, and the actuator is located in the crawl space below, drill a small, discrete, hole in the floor (e.g. inside the vanity where the wire will be hidden) and feed the wire up into the room.

4. PREPARE AND MOUNT THE SWITCH BOX.

Choose a location for the switch box. This should be easily accessible and near the outlet where the transformer will be plugged in. (If your switch box will be located outside or in a damp location use a waterproof switch box.) Remove the cover from the switch box. Drill a hole in the middle of the cover, just large enough so the toggle switch fits through. Insert the toggle switch through the hole and secure it using the toggle switch's locknut. Thread the ¾" × ½" reducing bushing into the lower knockout of the switch box, then attach the strain relief cord connector to the bushing. Note: If your box doesn't have threaded knockout holes use a suitable strain relief cord connector that fits with your box. Finally, mount the box to the wall with screws. Don't put the cover on yet because you'll be connecting the wiring inside the box.

In-wall option: If you want the switch to be flush with the wall (not in a box mounted to the wall), you can do this with a little extra work. First, you'll cut out a small section of the drywall where you want the switch to be located and then fish the wires up through the wall. Then, you'll attach a single low-voltage drywall mud ring to the wall and make the connections to the switch inside the wall. When you attach the wall-plate to the mud ring it will sit flush on your wall (you will also need to drill a hole to fit the switch as mentioned above). Note: Many states allow low-voltage wires to be run inside the wall, but be sure to check your local regulations about any potential permit requirements. If you're not familiar with fishing wires you'll need to seek out further resources for proper technique and the tools required; see Resources.

5. WIRE THE ACTUATOR TO THE TRANSFORMER AND SWITCH.

The final wiring connections will vary based on the type of actuator and transformer you have. A sample wiring configuration is shown below and on the next page for general reference. Once you've made all the wire connections inside the box, confirm that the actuator switch is turned on for testing.

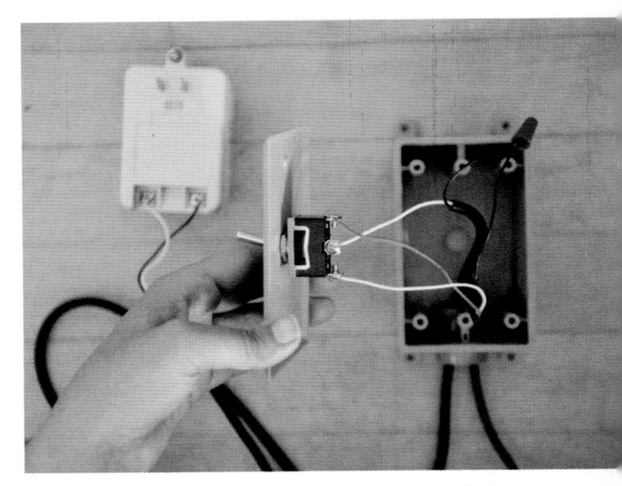

Step 5. Transformer wired to actuator and switch

ACTUATOR INSTALLED

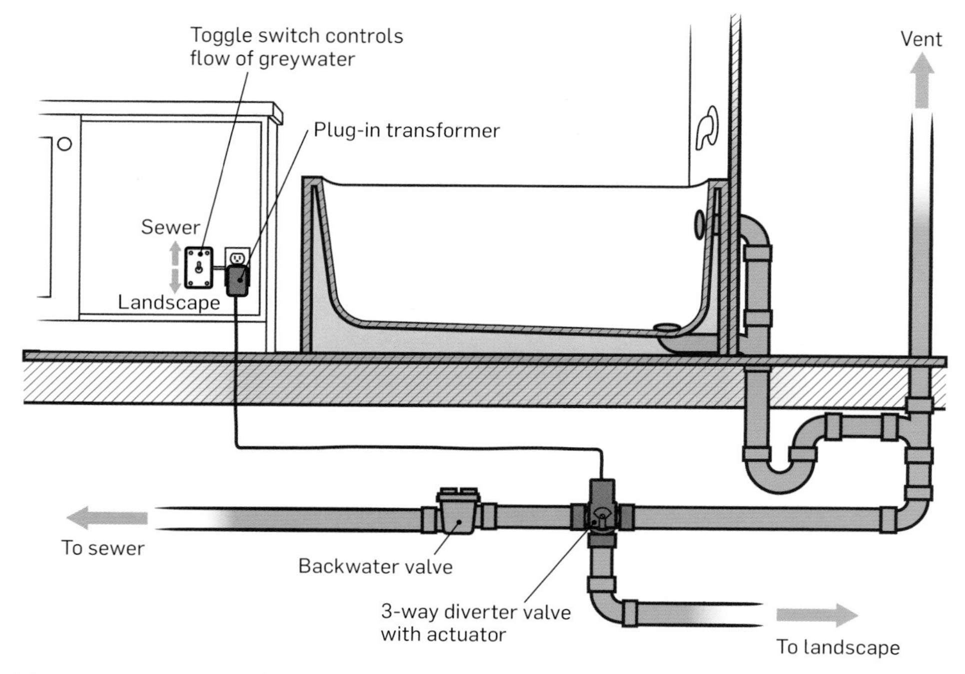

ACTUATOR WIRING (FOR GENERAL REFERENCE ONLY)

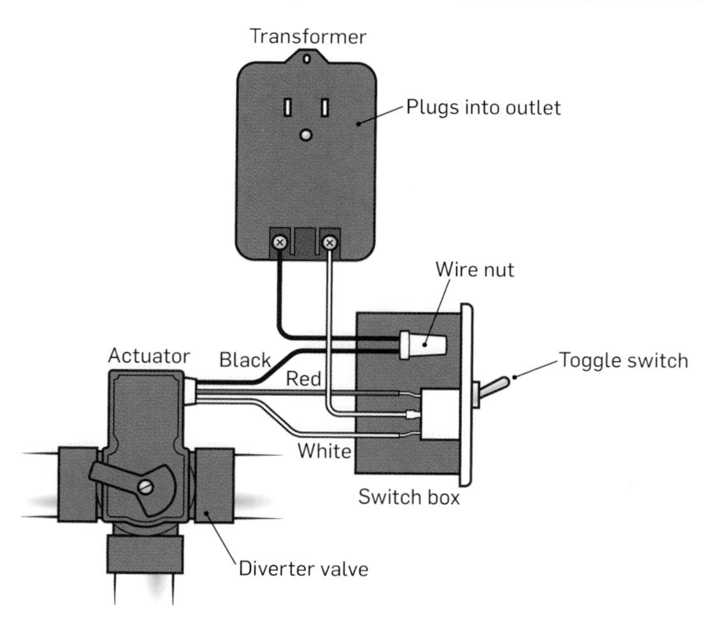

6. TEST THE ACTUATOR.

Turn the toggle switch up; the actuator should slowly turn the valve, stopping to shut off the desired port. Turn the switch down (if you used an on-off-on switch it will click through the center location before it clicks again to engage the motor). The actuator should now turn the valve slowly in the opposite direction. Note where the actuator turns the valve; for example, up = greywater, and down = sewer/septic (or vice versa). If the actuator doesn't turn the valve, first check to make sure it's turned on, then check the wiring. Lastly, adjust the cams (see step 1) if the valve is not turning appropriately.

When you've confirmed that the actuator turns the valve correctly, attach the faceplate to the switch box. Label the switch and secure the actuator and transformer wire as needed (using wire clips). Remember to unplug the transformer if you need to change the wiring.

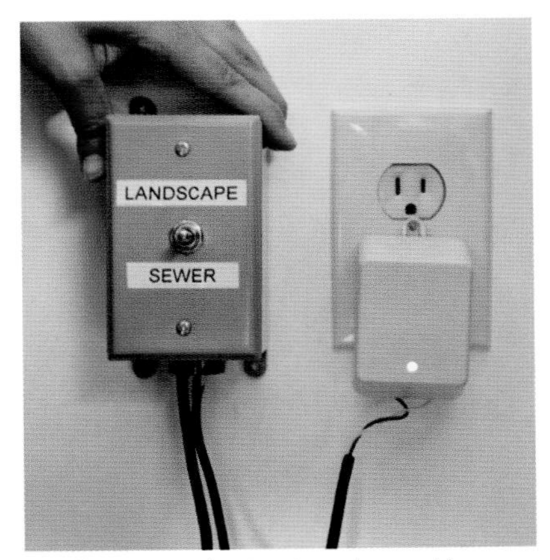

Wall-mounted actuator, plugged in

Example of flush mounted actuator switch

CHAPTER 10

Pumped and Manufactured Greywater Systems

Can't install a gravity system? Pumps can easily send greywater uphill to your plants.

Here's how: Greywater is directed to a small tank, called a surge tank, and then pumped to the landscape. The water is not stored for more than a day — 24 hours typically is the longest you're allowed to store it legally, and stored greywater stinks! The system requires access to the plumbing to install a diverter valve, room for the tank, and an electrical outlet to plug in the pump. Permitted systems may also need an electrical permit if the system requires a new outlet.

Pumped systems offer a few advantages over gravity-based systems. They can send water uphill and across long distances. They're also able to distribute water to more plants than gravity-based systems, and they're effective for combining flows and distributing the water around the landscape. As for the disadvantages, pumped systems use energy, and the pump will eventually need replacing, perhaps every 10 years if you're lucky. You can use a pumped system for drip irrigation, but you have to filter the greywater and use special greywater-compatible drip irrigation tubing; otherwise, the emitters will clog.

In this chapter I'll take you through the design and construction steps of a simple pumped system and discuss options for purchasing a manufactured greywater system that filters the water for drip irrigation.

IN THIS CHAPTER:

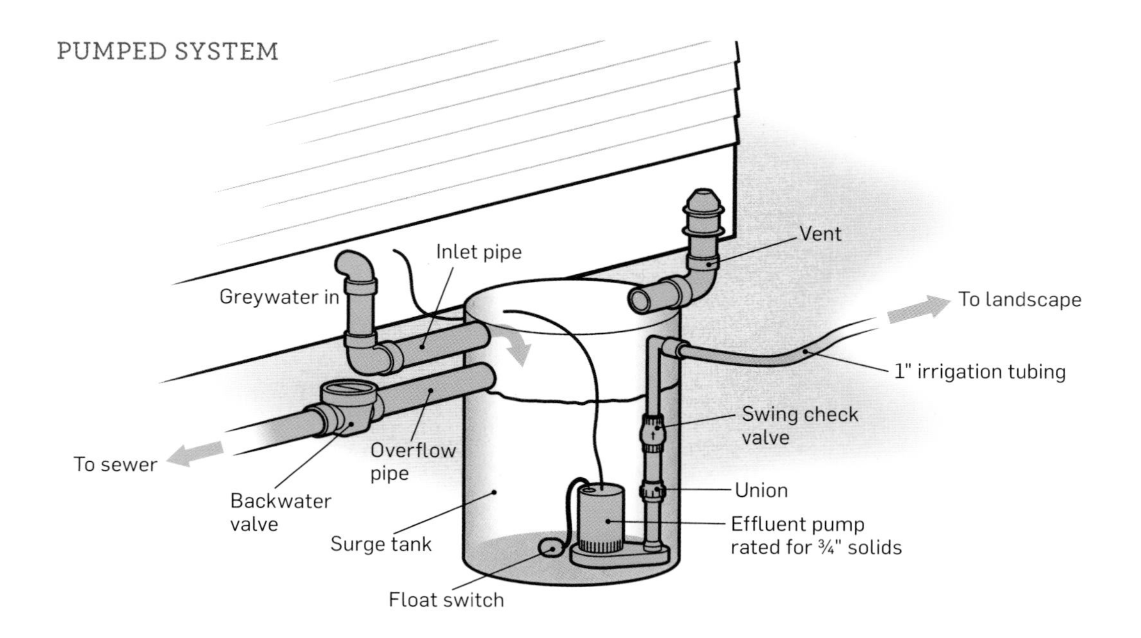

Design Considerations

The first question to ask yourself when designing a pumped greywater system is, "Does this system truly need a pump?" Many people assume that a greywater system needs a pump, when many don't. These situations, however, do require a pump:

- The only landscaped area needing irrigation is uphill of the greywater sources.
- There is a long flat area of hardscape that must be crossed before reaching the landscape.
- The yard is flat and the only irrigation needs are far from the house.
- The landscape requires drip irrigation (e.g., lots of small plants).

The first step in designing a pumped system is to identify where to install the 3-way diverter valve. See Diverter Valve and Pipe Runs on page 122 for details.

A pump can send greywater to any part of your landscape, even into separate irrigation zones. Try to route as many greywater sources as you can into the tank. Tank capacity depends on how many fixtures flow into it; size it so there is enough room for all the water to enter and be pumped out without overflow. This is typically 30 to 100 gallons, though the manufactured system Grey Flow uses a smaller collection chamber before pumping out the water (see pages 157–158).

The tank needs an overflow drain in case the electricity goes out or the pump breaks. Ideally, the drain overflows into the sewer/

septic and alerts you of any system failures, such as the pump going out. If your project requires a permit, an overflow to the sewer/septic likely will be mandatory. (Although, conventional sewage ejection pump systems, nearly identical to this system, do not have an overflow and are widely permitted.) Sometimes it's not possible to overflow to the sewer/septic, in which case you can overflow into a nearby mulched area or skip the overflow and find out that something is wrong when the shower stops draining.

Where to Locate the Tank?

Common tank locations are adjacent to the house, in a basement, buried outside the house, or buried in a crawl space. Some state codes have setback requirements for the tank location, so talk to your local permitting authority for details. Following are three basic setups.

HOUSE WITH A BASEMENT: Plumb greywater sources into the tank and locate the tank near an electrical outlet.

HOUSE WITH A CRAWL SPACE: Place the tank outside of the house. If overflow to sewer is impractical, create an overflow to a mulched area near the tank.

HOUSE WITH A VERY LOW CRAWL SPACE: Either bury the tank under the house or, if you can't fit a tank into the crawl space, core-drill through the foundation to get the pipe outside and then bury the tank near the house.

If you get a permit for a pumped system, there may be additional requirements not included in the following instructions. Talk to your inspector early on to find out exactly what's required for your project. Some jurisdictions require backflow prevention with any pumped system — even those with no potable connection — to prevent problems due to future alterations to the system that may create a cross-connection. Installing a backflow prevention device will add $500 to $1,000 to the cost of your system, in addition to a potential annual inspection fee of the backflow prevention device.

TANKLESS PUMPED SYSTEM OPTION

The company SaniFlo makes several models of effluent pump units for shower, sink, and washing machine greywater. The pump is designed to move greywater up to a sewer line from a basement level room, which makes it compatible with pumping greywater out of a crawl space and across a flat landscape or up a small slope. Some greywater installers opt for using SaniFlo effluent pumps instead of putting together their own tank and pumped system because the SaniFlo systems can be cheaper, easier, and faster to install. All these systems require venting. Greywater is pumped out to the landscape using the same L2L-type irrigation system (see page 154 for irrigation instruction details).

Building a PUMPED SYSTEM

This project provides an overview of pumped system construction. There are many variables that impact each specific installation, including the existing plumbing, tank location, permit requirements, and the landscape. The installation of the diverter valve can be identical to that of a gravity branched drain system, while the tank location determines where you plumb the greywater pipe and the overflow. Due to the variability of pumped systems, I'll cover just the basic process to give you a sense of what's involved; specific construction details will be dictated by your project requirements.

A few common challenges you may encounter when installing your pumped system are installing a below-grade tank, working in a small crawl space, tapping into old plumbing pipes, and (for projects requiring a permit) stringent permit requirements. For example, an inspector may ask for a drain on the tank, but if the tank is below grade, installing a drain is nearly impossible. Once the plumbing and tank are installed, the irrigation portion is much simpler, faster, and lower-cost (in materials) than for a branched drain system.

1. INSTALL A 3-WAY DIVERTER VALVE.

Follow the same procedures as when installing a diverter for a gravity-based branched drain system (see page 128). This also includes the option of installing an actuator to control the valve remotely (see page 142).

2. INSTALL THE SURGE TANK.

Surge tanks typically are 30 to 50 gallons for a single fixture, and 50 to 100 gallons for multiple fixtures. The surge tank must be near an outlet so you can plug in the pump. If you do not have solid, flat ground, you'll need to make a gravel or concrete pad for the tank. Some codes require concrete pads. If your surge tank is below grade make sure it is protected from uplift; if the water table rises, an empty tank will literally float out of the ground. Many buriable, pre-fabricated sump basins come with an "anti-floatation device," but if you use a repurposed tank and the water table can rise enough to reach the tank, you'll need to anchor it yourself. For example, construct a bottom hold-down with a slab of concrete below the tank and attach the tank to it, or seek out other suitable methods. Be sure to label your surge tank: *Caution Non-Potable Greywater Do Not Drink* (or whatever your state code requires).

HOW TO CHOOSE A PUMP

A high-quality pump will last many years. Look for an **effluent** pump that's rated to pump ¾" solids (so it can pump out anything that gets down the drain), is fully submersible, and operates on 115/120-volt (not 230/240-volt) electrical power. A typical sump pump is not powerful enough for this system. Some people use smaller-sized pumps, rated to pump ⅜" or ½" solids, though these aren't as strong as ¾"-rated units. Note that most effluent pumps are designed for a larger discharge pipe than the 1" irrigation line used in this system. I'm not aware of any problems in the field with reducing the line to 1", though some pump companies may recommend against it.

Because this system is not the standard application for a pump, it can be harder to get help from a pump specialist. Irrigation system pumps typically are sized based on the combined factors of how high and far they will be pumping, as well as the pressure and gallons per minute (gpm) required by the irrigation system; for example, 30 psi at 15 gpm. Since unfiltered greywater doesn't have these pressure requirements, typical pump sizing calculations do not apply. In most situations any pump rated for ¾" solids will be more than powerful enough to irrigate a typical yard. If you'll be pumping up more than around 20 feet you'll need a beefier pump than for a less steeply sloped yard. Another option is to use a SaniSwift pump, which is preinstalled in a container and easy to hook up; see page 150 for details.

3. PLUMB THE GREYWATER PIPING INTO THE TANK.

Identify where the greywater source(s) will enter the tank. Drill the appropriate size of hole in the top of the tank (not the lid) for the watertight fitting. See Tapered vs. Straight Threads, page 179, for options for a watertight connection (either a bulkhead fitting, Uniseal gaskets, or electrical conduit male adapter with an electrical female adapter and a washer). Use a hole saw to drill a clean hole.

Add a union before the pipe enters the tank so it's easy to disconnect the pipe. Plumb the pipe into the tank. Do not screen the inlet or outlets. Screens will quickly clog and require frequent cleaning.

Alternatively, you can purchase a sewage basin made with preformed holes. Called "knock-outs," these can be conveniently punched out, so no drilling is necessary.

4. PLUMB THE OVERFLOW.

Overflows connected to the sewer/septic must be equipped with a backwater valve in the horizontal position to prevent the possibility of sewage backing up into your tank and getting pumped out to the garden. Locate the overflow slightly lower than the greywater inlet. Install a bulkhead-type fitting the same size as used for the inlet. If you plumbed more than one pipe into the tank, size the overflow to accommodate the total inflow.

Plumb the overflow either back to the sewer (for permitted systems and any system where this is possible) or to a mulch basin. Install a tee-type fitting to connect the overflow to the sewer pipe; the specific fitting will depend on your plumbing configuration. Install the backwater valve. An overflow to a mulch basin doesn't need a backwater valve here because there is no sewer connection.

5. INSTALL A VENT.

The tank must be vented. You can use a standard plumbing vent or an AAV (see page 179). Alternatively, if the tank isn't connected to the sewer (and doesn't require a permit), any loose seal will let air into the tank and function as a vent. If you connect to the household plumbing vent system, make sure you are above the highest sewer connection so a clog couldn't backflow sewage down the vent and into your tank.

6. INSTALL THE PUMP.

See How to Choose a Pump (page 152) for help with pump selection. Place the pump in the tank and make sure the float switch can move freely; if it gets hung up on the tank wall, the pump won't turn on (or off). Drill a hole for the pump's cable, above the inlet and overflow. Use a waterproof cord connector if your system requires a sealed connection.

Run 1" PVC pipe from the pump to the outside of the building. You'll need a few fittings to connect the pump to the 1" pipe, and these depend on the type of pump you have. If your landscape is above the tank, install a 1" *swing check valve* in the 1" pipe in the tank. This prevents water from flowing back into the tank; if enough water flowed back in, it could cause the pump to cycle unnecessarily.

Adjust the float switch so the smallest amount of water is allowed to sit in the tank; that is, the pump stays on as long as possible.

If you are adding a filter to the system it would be installed here, after the pump and before the irrigation system.

Tank interior showing effluent pump and inlet and overflow pipes

Installing a pumped system: pump in tank with piping installed

7. INSTALL THE IRRIGATION SYSTEM.

The irrigation portion of a pumped system can be identical to that of a laundry-to-landscape system (page 80), with a couple of key differences. First, you can supply more outlets (20 or more) with a pumped system, although you'll need to consider how much water stays in the line if you have very long runs and a small surge tank. Second, you can cap the end of the main lines on a pumped system, and you'll clamp the tee connections on the 1" tubing (these can come apart with the higher pressure from the effluent pump). You can also have multiple zones in a pumped system; see Irrigating with Multiple Zones (below) for ideas, keeping in mind that one zone is simplest. Zones that are controlled manually will require you to switch between them on a regular basis, while automatically controlled zones will add cost and complexity to your system.

On steep slopes be sure to run the tubing to the highest irrigation point, then come down the slope to irrigate the remaining landscape. And, if your site has a lot of elevation changes, you can use zones to help distribute the water.

IRRIGATING WITH MULTIPLE ZONES

Standard timers, controllers, and irrigation valves used to create multi-zoned irrigation systems with potable water are not compatible with a simple greywater system. Often, thoughtful planning of the greywater system — and successfully directing different fixtures to distinct areas of the landscape — can obviate the need for multiple zones. However, there are several situations when you may want to spend the extra time and money to create a multi-zoned greywater irrigation system:

- You produce more greywater than you can logistically spread out over one area.
- You want to fully dry the soil in between irrigation periods.
- You want some areas to receive more water than others.

Greywater can alternate between zones by manual or automatic methods. The simplest method is manual: Add a second 3-way valve outside, in an accessible location. This creates two zones for which you manually turn the valve to switch back and forth (see page 118).

Before choosing a manual method, ask yourself whether it's realistic that you will switch this valve every week for the next 10 years. If not, install a single zone or an automatic switching method. One automatic option to create two zones with unfiltered greywater is to use a Jandy or Pentair type valve with an electronic actuator, controlled by a 7-day timer. Program the timer to switch between the two zones every few days (the cost for this is around $250).

Another common method for automatically controlled zones is to employ an indexing valve that distributes greywater to multiple zones

Remember to use mulch basins in the landscape to soak up greywater and slowly release it to the roots of plants, and to size the system based on your estimated greywater production and the plant water requirements of the landscape (see chapter 3 and How Much Water do My Plants Want? on page 49). Your basins will probably be a lot smaller than those for a branched drain or laundry-to-landscape system because less water will come out of each outlet. You'll also be able to use the smallest mulch shield options (sections of 4" drain pipe; see page 115).

8. TEST AND MAINTAIN THE SYSTEM.

Plug in the pump and run water through the system. Check for leaks and confirm that the pump turns on/off appropriately. Also check for even water distribution in the landscape. If necessary, fine-tune system as described for an L2L system (see page 107).

Perform the same regular maintenance on the irrigation lines as with an L2L system. In addition, check on the tank and pump annually, or as needed. Clean out the tank if the sludge layer builds up too much.

(up to eight). Each time the pump turns on, the indexing valve rotates to a new zone, without need for a controller or power supply. Many indexing valves require high pressure to operate; try a lower-pressure one, such as Fimco's *Wastewater Hydro Indexing Valve* (10 psi; costs about $90), which requires minimal filtration. You may need to include a filter to use this valve. Orenco makes a mechanical distribution valve (about $270) that operates with their compatible pump (see Resources for manufacturer websites). While this valve is designed to require filtration (80-mesh), installers report that it works without any. K-Rain also makes an indexing valve designed to work with dirty water (25 psi; costs about $80).

Another option is the Grey Flow Rotor Valve ($160), a volume-based indexing valve, which means each zone receives an equal amount of water (13 gallons); very handy for planning irrigation requirements. The valve requires at least 3-gallons-per-minute flow rate, and 40-mesh filtration. The valve comes with six outlets, and you can combine them if you want fewer.

INDEXING VALVE

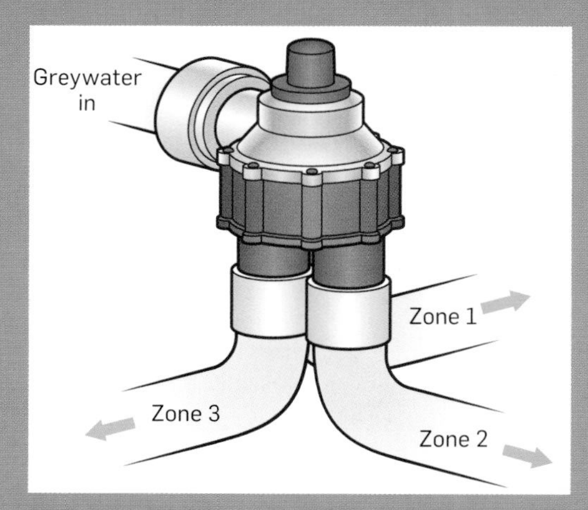

Each time the pump turns on, the valve rotates to a new zone. (It requires an appropriate pump for the valve to function properly.) Indexing valves typically have two to eight zones.

Manufactured Greywater Systems

The holy grail of greywater systems: a low-cost "kit" that sends greywater into drip irrigation tubing without the need for regular maintenance. It works in every home and landscape. And it won't break in a year. Manufactured greywater systems attempt to do this, with varying degrees of success. The greatest challenges come from the fact that greywater systems are not one-size-fits-all, and they're a hybrid of plumbing and irrigation systems. You won't find plumbers or landscapers buying a "kit" when they plumb a house or install an irrigation system; there are too many variables for that to be feasible.

Manufactured Greywater Kits: Pros, Cons, and Costs

If you need a pump, or want drip irrigation, manufactured systems offer a convenient package with the necessary components: tank, pump, and filter. In addition, you can get technical support from the sales representative for help with installing the system.

The main drawback with kits is that typically you end up with relatively low-quality parts and more frequent maintenance. These kits use parts made for other industries (there are no greywater-specific parts in North America), such as pool, pond, wastewater, or septic effluent equipment. Most systems require regular filter cleaning — an unpleasant task that if forgotten causes a system failure. Find out specifics about required maintenance on filters if you're considering installing this type of system. Or, pocketbook permitting, get a maintenance contract with an installer. Manufactured systems typically can't handle kitchen sink water, it will gunk up the filters.

The lower-cost greywater kits range from $500 to $800 and the mid-range kits cost around $2,000, not including the diverter valve; plumbing parts needed to plumb greywater to the system; irrigation components; or any associated preparation, labor, or permitting fees. Now we'll take a look at three systems currently on the market (see Resources for online suppliers). I'll discuss other manufactured systems for new construction in the next chapter.

AQUA2USE

This unit is housed in a small (approximately 2 × 1 × 2-foot) plastic box that has a series of filters (originally made for filtering fish pond water) and a low-capacity pump. It's rated to pump up to 16 feet at 14 gpm. Greywater is filtered to pass through a ¼-inch tube, so you could install an irrigation system with a ½-inch main line and ¼-inch branch lines or outlets to plants. If you want to use in-line drip irrigation, an additional 100-micron drip irrigation filter will be needed on the pump output. Use a disc-type filter, as these require less frequent maintenance than screen-type filters.

PROS OF AQUA2USE

- Filtered greywater can be spread out over a larger area than unfiltered greywater.
- The system is preassembled and includes the tank, pump, and filter all in one package.

CONS OF AQUA2USE

- Pump is small and not robust.
- Float switch can get hung up on tank edge and cause pump to cycle endlessly.
- Greywater storage capacity is small; when it gets backed up, greywater overflows to sewer and is wasted.
- Filters must be cleaned manually, about every 6 months; any disc filter on this system will require inspection every month.
- A clogged filter makes all greywater flow to the sewer until the filter is cleaned; plants won't be getting water when the filter is clogged.
- Kit doesn't come with any irrigation components.
- Greywater is not filtered enough to be sent directly into drip irrigation tubing (it will clog the tubing over time). Additional filters (not included in kit) are needed.

GREYFLOW

The Australian company Advanced Waste Water Systems (AWWS) makes several GreyFlow greywater irrigation systems for both residential and commercial applications. Now available in the United States, these systems can be incorporated into new construction or as part of a retrofit.

G-FLOW is GreyFlow's most basic system, with manually cleaned filters. Greywater flows into the system, though the filters, and is pumped out into the greywater-compatible drip irrigation system. The advantage to this system over the Aqua2Use is the filters are housed in a plastic trough, so when you remove them for cleaning, crud won't fall into the system and need to be removed. The disadvantage to this system is the filters must be cleaned manually.

GREYFLOW PS PLUG & PLAY features automatic filter cleaning. Greywater flows into the system, through the filter, and is pumped out into the irrigation system. On a regular basis the system automatically cleans the filter by blasting compressed air up through the filter, dislodging dirt and debris, and the incoming greywater washes it to the sewer. Using compressed air and greywater to clean the filter reduces installation complexity and permitting fees since there is no potable water connection. Other whole-house systems, such as the Gray-It (see page 158), use potable water

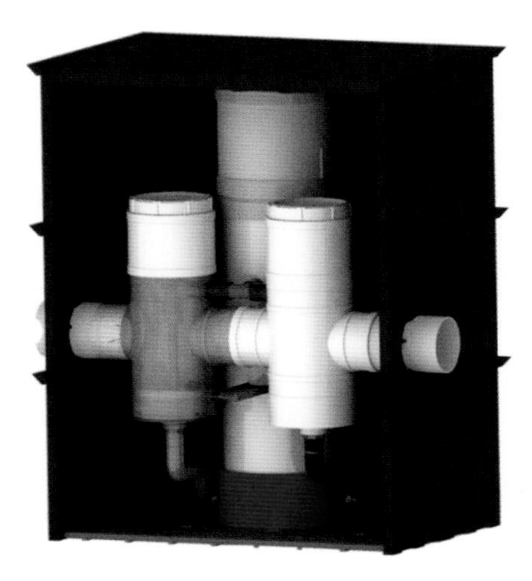

GreyFlow PS Plug & Play: the most affordable automatically cleaned filter system currently available

to flush the filters. GreyFlow systems can have up to six irrigation zones. The system dimensions are about 2½ feet deep × 2 feet wide, and 2 or 3 feet tall (a short or tall model).

PROS OF GREYFLOW PS

- Self-cleaning filter without a connection to potable water
- Greywater compatible drip irrigation, with multi-zones options (using a rotor valve, non-electric)
- System can be adapted to above or below ground installations
- The system is preassembled and includes the chamber, pump, and filters in one package

CONS OF GREYFLOW PS

- Relies on pumps and electricity

GRAY-IT

Gray-It is a home greywater system developed by the Israeli company Green Solutions. Greywater flows through a small filter, into the pump tank, where it's pumped out through a disk filter into an irrigation system. All filters require manual cleaning. The irrigation controller monitors how much greywater goes through the system, and if the irrigation program is complete any additional greywater generated that day will flow to the sewer. If there isn't enough greywater generated the system will bring in makeup water to complete the irrigation cycle.

The controller will show an error message if the system was unable to complete its schedule, which can alert the system owner to problems (if they happen to go outside to look at the controller). Gray-It doesn't include the irrigation portion of the system.

PROS OF GRAY-IT

- Uses efficient drip irrigation
- Controller monitors greywater flows and displays any problems

CONS OF GRAY-IT

- Relies on pumps and electricity
- Requires a potable water connection
- Manually cleaned filters
- Manufacturer recommends adding a pre-filter if using laundry water

CHAPTER 11

Other Types of Greywater Systems

In this chapter I'll summarize a handful of additional types of greywater systems, some of which are simple enough to build yourself. Others are more complex and require professional assistance.

Greywater systems that aren't connected to a house, such as those drawing from an outdoor sink or shower, can be extremely simple and functional. Some systems have specialized functions, such as ecological disposal of greywater or supplying year-long irrigation in freezing climates by incorporating an indoor greenhouse. If you're planning to build a home and want a more advanced system, there are high-end systems that collect, filter, and distribute the water and offer remote monitoring capability.

IN THIS CHAPTER:

Outdoor Fixtures

Have an outdoor shower, sink, drinking fountain, or washing machine? They're perfect opportunities to capture and reuse greywater! Greywater sources located outside make for a simple system, just directing water to nearby plantings. I'll describe two common systems here, and you can adapt the ideas for any outdoor water source.

This outdoor shower drains to irrigate nearby plants.

Outdoor Shower

Simple and easy to install, an outdoor shower is a perfect way to irrigate nearby plants. Imagine a cool shower on a hot day, water flowing over your body and draining directly to nearby plants that adorn the shower structure. When you build the floor of the shower, use flagstone or a small concrete pad and slope it toward a well-mulched planted area adjacent to the shower — no need for a drain or plumbing. For elevated floors, slope the soil below and cover it with rock to move water toward the plants. If you don't want to plumb water lines to the shower, use a food-grade garden hose to supply the water. For a warm shower, use coils of black irrigation tubing to heat the water passively, through sun exposure.

Outdoor Sink or Drinking Fountain

Every community garden or outdoor seating area needs a sink for washing garden produce or dirty hands and for supplying drinking water. Greywater from the sink can irrigate nearby plants via a small branched drain system. Supplying the sink from an existing spigot means the fixture isn't permanently connected to a building's plumbing system and therefore won't technically produce "greywater" or be illegal without a permit . . . in most cases; the legality of this is a "gray" area. Obviously it's not illegal to wash your hands using a garden hose and let the water run on the ground, but if you build a small structure and install a sink using the garden hose, will that produce greywater and require a permit? This legal ambiguity makes these systems attractive in schools or community

This outdoor shower directly irrigates the landscape via greywater runoff from its stone floor.

centers where permitting greywater systems may be particularly challenging.

Whole-House Greywater Systems

There are several greywater systems designed for use in new construction (or major remodels), when you are able to capture and reuse all greywater sources in the home. Most of these systems are relatively new, and relatively expensive, but they allow a home to recycle all its greywater and employ an automatic drip irrigation system capable of irrigating any type of landscape. Some of these systems treat the water for indoor toilet flushing as well. Even with filtration and treatment of greywater, these systems do not remove salts or boron, which could harm plants if they build up in the soil, so it's still important to use plant-friendly products in the home. In terms of cost, the total installation cost is typically two to four times more than the cost of the system itself; it takes a lot of time and effort to get permits and install these complex systems. They all employ self-cleaning filters but will still require annual maintenance. Below, I'll discuss three types of systems currently available.

If the cost or complexity of these systems seem too much for your project, consider the GreyFlow PS (page 157), which is also suitable for whole-house greywater. It costs less than the other whole-house systems, and it doesn't have a potable water connection, which makes permitting easier. The GreyFlow system has a self-cleaning filter, drip irrigation, and multiple irrigation zones (without need for a controller).

Nexus eWater

This Home Water and Energy Recycling System is a fully automated whole-home option that integrates into the plumbing infrastructure of the house. It collects, filters, and disinfects household greywater for reuse, either for irrigation or toilet flushing. It is the first residential greywater system to receive NSF-350 certification for greywater, which in California allows the treated water to be used for toilets and spray irrigation. Additionally, the system can capture heat from greywater using heat pump technology.

Nexus estimates that its system can reduce home water consumption by 34% and reduce energy needed to heat water by 75%. In new homes, the total cost of installing the system is less than $10,000, with an additional $3,000 to add the energy recycling option. Annual filter and UV bulb replacement costs are $150.

(Water ReNu) IrriGRAY

This automatic irrigation system uses greywater, rainwater, condensate water, or other non-potable on-site water. (Note that this automatic system is the new focus for Australian (now US resident) greywater designer Paul James. His previous IrriGRAY system used a manually cleaned filter and is now discontinued.) The new IrriGRAY consists of a small sump basin, which temporarily collects greywater and pumps it out through a self-cleaning filter. The filter is automatically backflushed with pressurized, potable water for cleaning (requiring backflow prevention) and directed to the irrigation system. The irrigation lines are also flushed with potable water to remove buildup in the lines.

IrriGRAY uses a multi-zoned smart irrigation system, which can be controlled and monitored remotely (on your phone, computer, or tablet) and will bring in makeup water if there isn't enough greywater to complete the irrigation cycle. The monitoring system notifies the owner if there are any problems, including leaks. IrriGray systems were installed in a subdivision in San Angelo, Texas, and used in both residential and commercial applications. The whole system, including irrigation parts, costs $5,000 to $7,000 for a typical single-family home.

ReWater

The ReWater system has been available for more than 20 years, far longer than any other greywater system currently on the market. It consists of a tank, a pump, a sand filter (with automatic backflush), and a sophisticated irrigation controller that sends greywater to specialized subsurface emitters. The filter is backflushed with pressurized potable water (with cross-connection protection), and the controller brings in makeup water at the end of the day if there isn't enough greywater to complete the irrigation program. The whole system, including irrigation components, costs around $6,000 to $10,000 for a residential landscape. ReWater Systems are typically installed in new-construction residential and commercial buildings.

PROS: WHOLE-HOUSE MANUFACTURED GREYWATER SYSTEMS

- Systems use drip irrigation, a very efficient way to irrigate.
- Can incorporate other on-site non-potable water sources into the greywater system, or use potable water, to replace other irrigation systems
- Fully automatic; filter is self-cleaning
- Able to spread greywater over a large area and irrigate any type of plant

CONS: WHOLE-HOUSE MANUFACTURED GREYWATER SYSTEMS

- Expensive and require a high skill level to install
- Relatively complicated; more moving parts that can break, as well as electronic components. Rely on pumps and electricity
- Most of the systems are relatively new and so don't have a long track-record of performance. (Even if the system designer or company is older, the specific system on the market may be relatively new.)
- Require backflow prevention so the system can't contaminate potable water supplies; this adds costs and permitting challenges.
- Filters don't remove salts or boron or other substances potentially harmful to plants, so plant-friendly products must be used in the home.

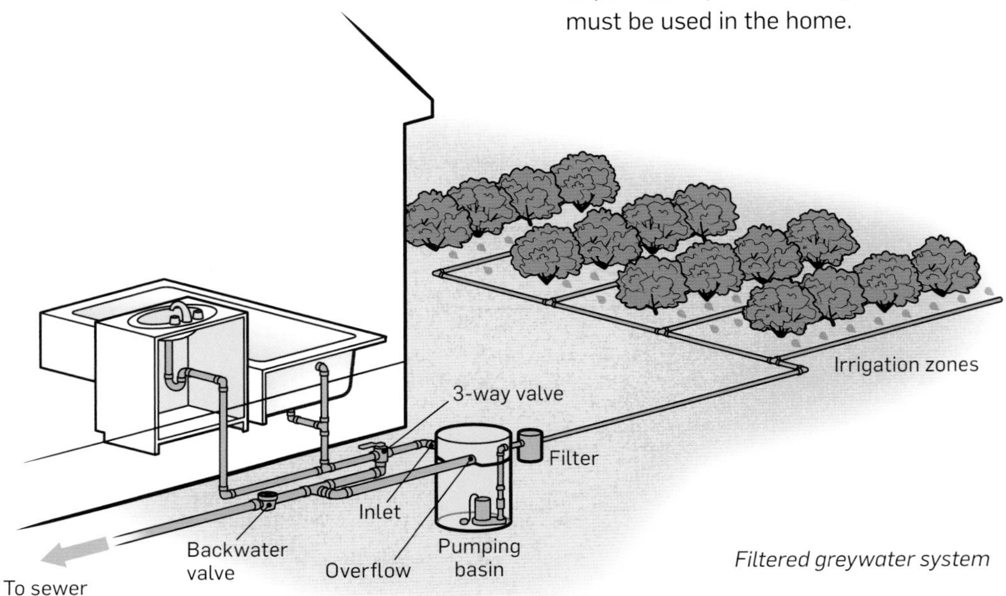

Filtered greywater system

Subsoil Infiltration Systems

Instead of mulch basins, subsoil systems use infiltration chambers to soak greywater into the ground. Greywater flows to belowground chambers designed to fill up with an inch or two of water, then soaks into the ground. Chambers are sized based upon estimated flows: larger chambers accommodate more water flowing into them. Subsoil systems are used where subsurface irrigation is needed and where a lot of water must infiltrate into a small area. They're also incorporated into some types of greywater greenhouses (see page 167). These systems often are installed to comply with local regulations and are commonly used in combination with composting toilets to replace the need for a septic system (for an example, see Sewerless Homes on page 166).

The cost of a professionally installed whole-house system ranges from $6,000 to $10,000 to replace a septic system. DIY installations using gravity flow to subsoil half-barrels may cost $200 to $300 in materials. There are a variety of materials available for subsurface irrigation:

INFILTRATORS. These are designed for septic systems. Use the shortest (in height) infiltrator possible so greywater will soak higher in the soil profile where there is more biological activity to process the water. The company Infiltrator Systems makes one that's 8" tall × 48" long × 34" wide (see Resources).

BOX TROUGH. Build your own infiltrators by creating a wooden box trough — essentially a rectangular box with no bottom. This allows you to customize the size and depth of the infiltrators to fit your site. Include an access hatch on top that opens for inspection and maintenance.

HALF-BARRELS. Use salvaged plastic barrels, usually 30-gallon-size, and cut them in half. Drill an entry hole on top and cut out an inspection hatch.

DOSING TANK. Professionally installed systems often use a dosing tank, which collects greywater and sends it all out to the infiltration area a few times a day. This allows the soil to re-oxygenate between doses. Dosing tanks can be pumped or work by gravity.

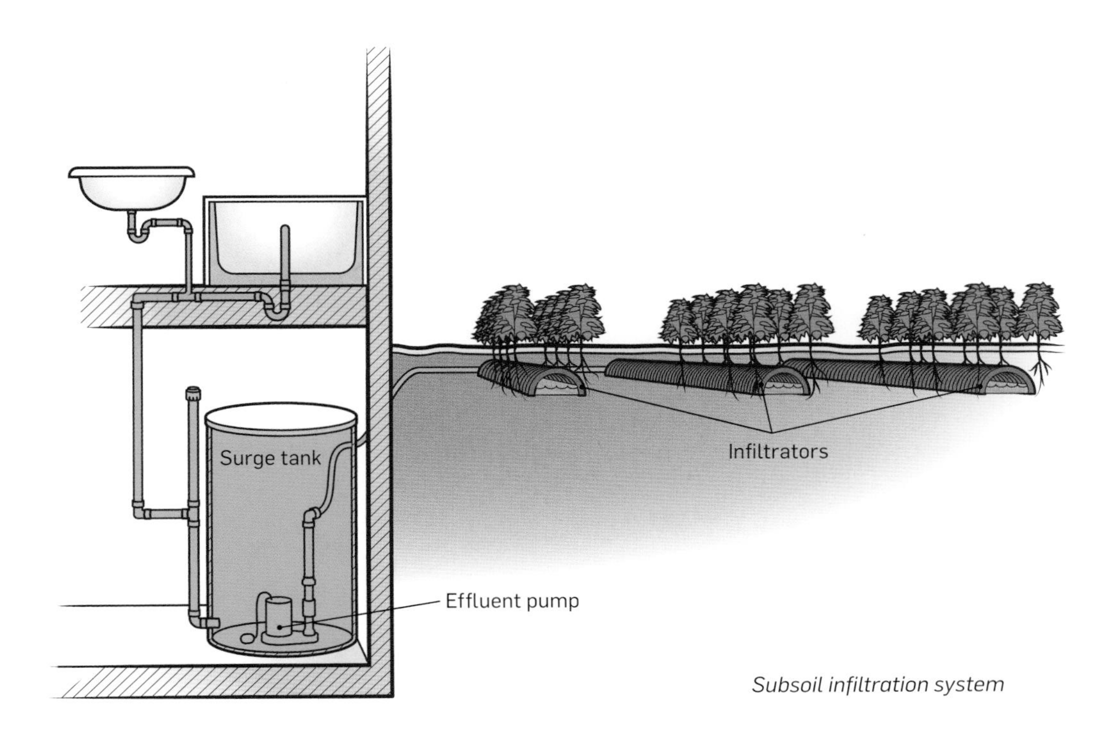

Subsoil infiltration system

PROS OF SUBSOIL INFILTRATION

- Keeps greywater below the ground surface, which is required by some state codes
- Able to send more water to a smaller area than with a branched drain system
- Meets code requirements for houses installing alternative wastewater systems, such as a composting toilet with greywater system
- Prevents groundwater pollution and recycles nutrients in upper levels of soil
- Professionally installed systems are ultra-low-maintenance

CONS OF SUBSOIL INFILTRATION

- Not a water-efficient method of distributing greywater; water is concentrated instead of being spread out to maximize its irrigation potential
- Requires a lot of digging and possible disturbance of existing landscaping

Sewerless Homes

Composting Toilet with Greywater System

John Hanson and his Maryland-based company NutriCycle Systems install whole-house systems (greywater with composting toilets) as an alternative to a septic or sewer system.

He also designs greywater systems used worldwide for the composting toilet company Clivus Multrum. His systems are maintenance-free: he has 10-year-old systems that require no maintenance whatsoever.

John's greywater systems satisfy health department concerns for approving no-septic homes or businesses (all sites must meet the percolation requirements of a conventional system for regulatory approval of an alternative one — this prevents development on a site unsuitable for a septic system). They are designed to prevent groundwater pollution and to recycle nutrients; they are not specifically for irrigation.

Greywater is delivered to subsoil infiltrators either with a pump or gravity dosing tank (non-electric). John uses Orenco's *automatic distributing valve* to create multiple zones in the pumped systems. His systems have no filters, and all organic matter decomposes in the soil. In freezing climates, pipes are buried deeper and never have standing water in the pipes. In most places, he infiltrates 8 inches deep, though goes deeper (12 to 16 inches) in areas with extreme weather conditions.

The cost for a residential greywater system ranges between $6,000 and $10,000. A Clivus Multrum composting toilet is $6,000 to $10,000. The total cost is between $15,000 and $20,000 — about the same as a new septic system.

John installed a NutriCycle system at his home in 1980, with two waterless toilets and one foam flush toilet (3 ounces per flush) connected to a Clivus Multrum toilet. The compost is removed once a year and used on the property. Household greywater is absorbed by a beautiful flower bed. You can visit John's website (see Resources) to see these systems in action and to contact him for an appointment.

John's advice to us: "Do it! Get involved, do something that is not polluting." He also advises public and commercial facilities to install these systems, since they are much cheaper than traditional on-site systems.

Greywater for Greenhouses

Greywater can be used to irrigate greenhouse plants. Common in cold climates, greenhouses extend the growing season and maximize the irrigation potential of greywater. There are two categories of greenhouses irrigated with greywater: standalone, outdoor greenhouses and integrated, indoor greenhouses that are attached to the house.

Standalone, outdoor greenhouses can be constructed anytime and the plants can be irrigated with a pumped greywater system. You can irrigate larger plants with a simple pumped system (no filter) or, to spread out the water to irrigate numerous small plants, include a filter for a drip irrigation system. Always cover the soil with mulch (either woodchips or straw) to catch particles in the greywater and prevent clogging of the soil. See Building a Pumped System (page 151) for help with system design and installation.

In some situations, a laundry-to-landscape system can also irrigate greenhouse plants. The washing machine must be close enough to the greenhouse (within 50 feet in a flat yard, or farther in a downward-sloping yard), and the plants must be either in the ground, in small raised beds, or in very large pots (wine-barrel sized). See chapter 8 for more information.

The cost to irrigate an existing outdoor greenhouse is the same as with a standard L2L, pumped, or filtered greywater system.

Integrated, indoor greenhouses irrigated with greywater are typically installed during new construction or a major remodel. **Earthships**, passive-solar houses made from natural and recycled materials, incorporate greywater greenhouses that grow beautiful plants and also filter the greywater to be reused to flush toilets.

In cold climates an indoor greenhouse requires proper siting to maximize sun exposure and minimize the need for supplemental heating. The cost for a professionally installed indoor greenhouse and greywater system may range between $10,000 and $30,000, on par with putting a small addition onto a home. The cost for the greywater system represents a small portion of the total cost for the greenhouse.

COMPOSTING AND GREYWATER AT CAMP

Bar-T Mountainside camp in Urbana, Maryland once had approval to put in a one-million-dollar septic system for their capacity of 350 daily campers. Instead, they installed a greywater and composting toilet system, saving more than $500,000 (even with an overdesigned greywater system and the cost of building a new basement for the compost chambers). Since the camped open in 2006 the system has helped to teach thousands of children about nutrient recycling.

PROS OF A GREENHOUSE SYSTEM

- Greywater can be used to irrigate all year long in cold climates (when outdoor plants are dormant).
- Indoor greenhouses facilitate growing tropical plants in just about any climate.
- Greenhouses can create zero-discharge systems, as the plants evapotranspire all the water.
- They can filter greywater for toilet flushing.

CONS OF A GREENHOUSE SYSTEM

- Building an indoor greenhouse is more complicated than simply installing a greywater system, and it requires proper siting (southern exposure in the northern hemisphere).
- Problems with indoor greenhouses can affect the living space (for example, a white fly infestation).
- Moisture levels and temperature of the greenhouse must be maintained.
- A pump may be required to move greywater into the greenhouse.

Indoor greywater greenhouse

Indoor Greywater Greenhouse
(with Composting Toilet)

Carl and Sara Warren run a residential design-build contracting company in eastern Massachusetts. When they decided to convert their barn into an office, they found out the septic system wasn't large enough for increased flows. They decided to install an alternative, *zero-discharge* system.

Composting toilets are an accepted technology and not a problem for permitting in their state. The Warrens chose a Phoenix composting toilet (Carl says they've had absolutely no problems with it and feel sorry for people who have to put up with the smell, water use, and cleaning hassles of a flush toilet). Getting a permit for a greywater system was harder and required approval from the state department of environmental protection; their system is monitored as a pilot project.

The system begins with water-efficient fixtures (washing machine, shower, and bathroom and kitchen sinks) draining to a surge tank. Greywater flows by gravity to ¾-inch PVC pipes spread throughout greenhouse planters. The planters are insulated and lined with an impermeable rubber membrane (required by the state department of environmental protection) and filled with a combination of peat moss, vermiculite, loam, and compost. Carl was surprised at how tolerant the plants are, and they experimented to find ones that thrived and evapotranspired the most water; bananas won. The beds transpire 50 gallons a day, even in the winter.

Maintenance includes trimming plants, annual cleaning of the grease interceptor (under the kitchen sink), biannual cleaning of the lint collector, and regular cleaning of strainers in the sink and shower drains.

Carl reflects that the greatest drawback to his system is the heating requirement. Because the greenhouse had to fit into an existing building, he wasn't able to orient it for passive heating. With proper orientation and an automatic insulating shutter system, he believes it could be 100 percent passively heated.

Cost:

- Greenhouse (12 × 21 feet): $20,000
- Nighttime shutter system: $3,000
- Greywater system: $700
- Composting toilet (Phoenix R-200, middle sized): $6,000

Constructed Wetlands

A **constructed wetland** is a watertight planter, typically lined with a pond liner, filled with gravel and planted with wetland plants. The plants, gravel, and microbes around the roots filter greywater and remove nutrients. More common in wastewater treatment plants and commercial-scale greywater than in backyard systems, constructed wetlands ecologically "dispose" of the water, instead of efficiently reusing it. Constructed wetlands are well suited for homes that produce more greywater than is needed in the landscape. And, importantly, in places without sewer treatment, constructed wetland systems treat household greywater to prevent water pollution.

Anyone researching constructed wetland systems will read about the importance of retention time: how long each water molecule remains in the system. In a municipal-scale system that treats wastewater for discharge into a waterway, it's critical to have sufficient retention time to ensure all the nutrients are removed from the water. In a backyard wetland, retention time is not important: the nutrients in greywater will fertilize the garden, and the water won't be discharged into a waterway. Backyard wetland system designers should focus on clogging prevention and surge capacity rather than retention time.

Backyard constructed wetlands typically are used for ecological disposal of greywater in climates with ample rainfall or places without sewer treatment. With my own system, I learned that I could easily grow the wetland plants I love without the drawbacks of flowing all the greywater through the wetland prior to the irrigation system by directing a portion of greywater to irrigate the wetland and using the rest for other plants. Costs for systems range from a few hundred dollars for a small do-it-yourself installation (or less if you use salvaged materials) to many thousands of dollars for a large, professionally installed system.

Here are a few things to keep in mind when designing a wetland system:

- How large? A very rough rule of thumb for sizing a wetland is ½ to 1 square foot per gallon (per day), with a depth of 1 to 2 feet.
- Use well-washed pea gravel for the substrate, and larger rocks around inlets and outlets.
- Use pond liner or an old bathtub to create a watertight planter. Sometimes wetlands are made out of cinder blocks or a strong wooden planter with a liner.
- Do not allow water to surface. Keep the outlet lower than the inlet to prevent surfacing greywater.
- Wetlands will require weeding and thinning of plants.
- Send the overflow to an appropriate location. For zero-discharge systems the overflow should be pumped back into the inlet of the wetland or any prior surge tank. Otherwise, send it to a well-mulched area.
- Don't plant fibrous-rooted plants near the inlet or outlet; they can clog up the system. Plant cattails, papyrus, equisetum, and canna lily or other plants well suited for your climate.

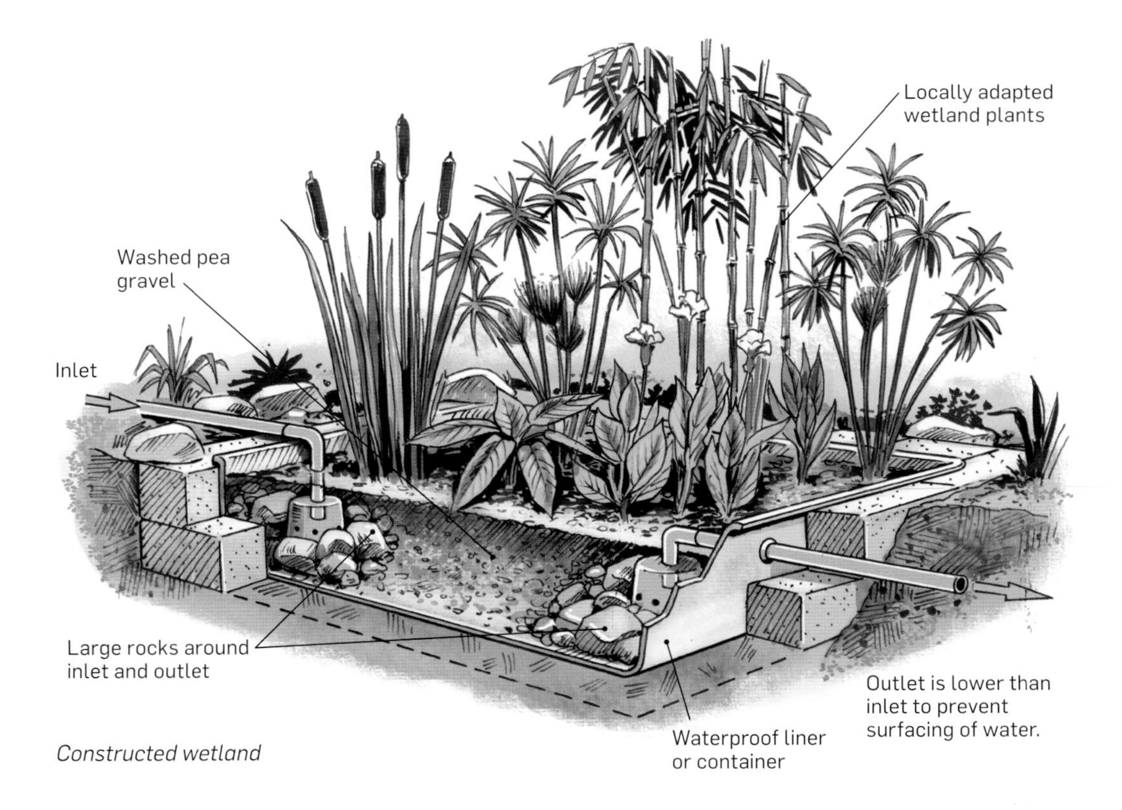

Constructed wetland

PROS OF CONSTRUCTED WETLAND SYSTEMS

- You can grow water-loving (and beautiful) wetland plants without using potable water.
- Plants and microbes remove nutrients from water, ecologically cleaning it.
- Water-loving plants consume lots of water, reducing the quantity of greywater to manage.

CONS OF CONSTRUCTED WETLAND SYSTEMS

- Less irrigation water is available for other plants if greywater passes through a wetland first.
- Clogging can occur in the wetland from overgrowth of roots that fill the air spaces in the gravel. This is time-consuming to remedy.
- If greywater contains salts, the wetland can increase their concentration. Wetland plants evapotranspire water but not salts, resulting in lower-quality irrigation water than incoming water.

Sand Filter to Drip Irrigation

Sand filters are used in both drinking water and wastewater treatment, and often there is confusion between the two processes. **Slow sand filters** clean water for drinking. Small quantities of non-potable water slowly drain through sand, where microbes remove pathogens and contaminates from the water. It is a biological process. **Rapid sand filters** treat wastewater. Greywater, for example, is pumped rapidly through a sand filter where the hair, lint, and gunk stick in the sand; filtered greywater comes out (not drinking-water quality!). Filtration is adequate for drip irrigation systems without clogging the small emitters. It is a physical process.

Rapid sand filters, like those used in swimming pool systems, are employed in some high-end greywater systems, like the ReWater System (see page 162). These systems use pumps, tanks, controllers, and drip irrigation and are much more expensive than other types. In general, a sand-filter-to-drip-irrigation system is installed in whole-house greywater systems, in high-end residential, multi-family, and commercial-scale new construction.

In a typical rapid-sand-filter system all greywater from the house is plumbed to a surge tank where the greywater is stored temporarily. Inside the tank an effluent pump, turned on by an irrigation controller, pumps greywater through the sand filter, then out to the landscape. The hair, lint, and other particles are filtered out by the sand, and greywater is distributed to plants via greywater-compatible irrigation tubing (made for greywater or septic effluent). If there is not enough greywater to complete the irrigation cycle, the system automatically supplements with domestic water. The controller automatically cleans the sand filter by pumping fresh water backwards through the sand, removing the lint, hair, and particles to the sewer. The cost of this type of system ranges from $10,000 to $30,000.

PROS OF A SAND FILTER SYSTEM

- Very efficient form of irrigation
- Replaces other irrigation systems, since greywater can be supplemented with other water sources to meet any irrigation need
- Fully automatic; filter is self-cleaning

CONS OF A SAND FILTER SYSTEM

- Expensive and requires a high skill level to install
- Relies on pumps and electricity
- Requires backflow prevention so the system can't accidentally contaminate potable water supply. This adds cost and permitting challenges.
- Filter doesn't remove salts, boron, or other substances potentially harmful to plants. It removes only large particles that would otherwise clog the irrigation system.

Greywater Goes High-Tech

Smart Systems Save Both Water and Time

A high-tech greywater system not only irrigates the plants, but it also knows how much water your home is using and reusing. All info is uploaded to a web page, where you can monitor the real-time water use (daily greywater flow and municipal use), view charts of monthly usage, and get email alerts for pipe breaks or leaks (this could save a lot of water!).

John Russell, owner of Water Sprout (see Resources), a design-build company specializing in greywater and rainwater systems, uses technology to automate and monitor his systems. John has fine-tuned his system over the past 10 years, using various filtering methods, all self-cleaning and operating with controls and makeup water.

He recently installed a system in a LEED-certified new home in Kentfield, California, collecting all household greywater (except kitchen sinks) for landscape irrigation and rainwater for reuse in toilets and laundry. Greywater flows into a 300-gallon underground tank for temporary storage. The system pumps filtered greywater to the landscape drip irrigation. The filters are automatically backflushed once a week to reduce system maintenance to once a year.

John has more experience with greywater filters than anyone I know. His recommendation to those considering filters:

"Filtering greywater appropriately is probably the most challenging aspect of utilizing greywater. When purchasing systems it's important to understand how often the filters need to be maintained, and try to choose systems that require minimal maintenance. Who wants to spend their weekend cleaning greywater filters?"

CASA DOMINGUEZ MULTIFAMILY GREYWATER SYSTEM

Casa Dominguez provides affordable housing to 70 families and transition-age youth exiting the foster care system as well as a child-care center and health clinic. Greywater from the laundry irrigates a beautiful courtyard and the perimeter landscaping. The "sand-filter-to-drip-irrigation" system was made by the company ReWater. Casa Dominguez is LEED Platinum certified, the highest level of certification from the U.S. Green Building Council, and obtained the first permit for a multi-family building to reuse greywater in Los Angeles County.

Reusing Septic Tank Effluent for Irrigation

Homes with a septic tank system may be able to reuse the septic effluent water for irrigation, with just a few alterations to the conventional system. A conventional septic system consists of a septic tank and a drain field, also called a leach field. Wastewater from the home flows into the buried septic tank. Solids in the water sink to the bottom and are decomposed by anaerobic bacteria while the liquids, called septic effluent, flow out the other end of the tank and into the leach field. Leach lines are made from large, perforated pipe buried in gravel-filled trenches. The effluent flows into the leach lines and out through the holes, and soaks down into the surrounding soil.

Homes with a septic system can adapt the tank to reuse all the effluent without separating out greywater flows. These systems treat septic effluent to irrigation quality and often are used when a traditional septic leach field is not suitable (for example, if the land is rocky and offers poor infiltration) or to capture irrigation water. Some states allow this treated effluent to be reused for irrigation. Most systems add oxygen to the septic tank to feed aerobic bacteria that clean the effluent to a higher quality. Companies such as Orenco make whole systems designed for septic effluent reuse. Other products, like the *SludgeHammer Aerobic Bacteria Generator* (see Resources), are installed inside an existing septic tank. Reed-bed constructed wetlands are also used to treat the septic effluent for irrigation.

In terms of cost, if the local authority allows the septic leach field to be reduced or eliminated, the cost is comparable to that of a traditional system. By contrast, regulatory requirements could make the system more expensive than a traditional one if they require a conventional leach field system in addition.

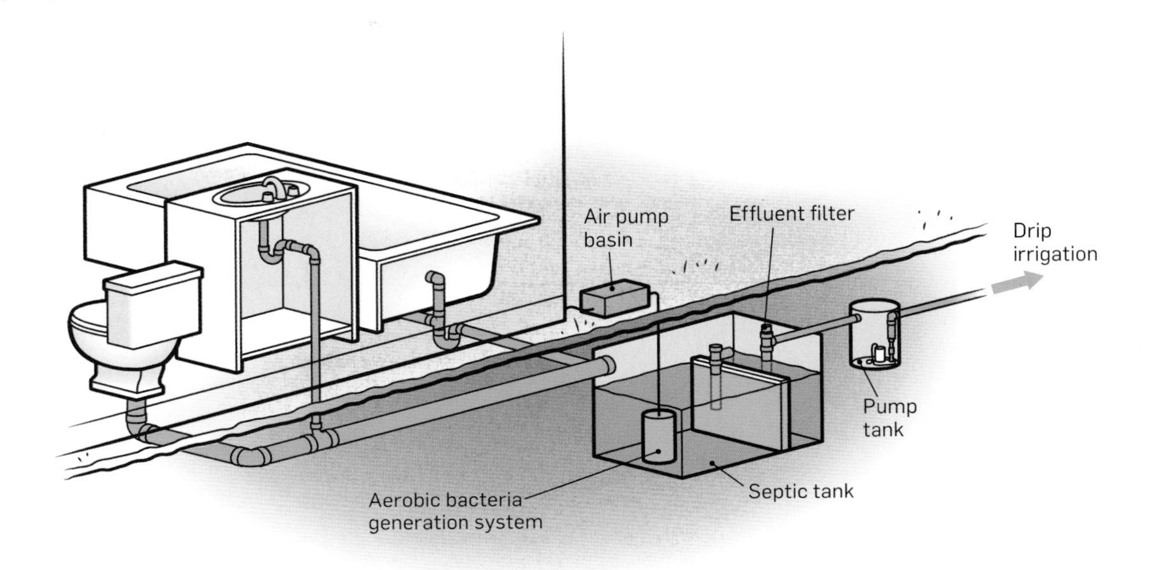

Septic tank effluent to irrigation

Typical costs for professional installation are $7,000 to $20,000. Materials only — for homeowner installations — run around $4,000.

PROS OF REUSING SEPTIC EFFLUENT

- Able to reuse all water, even from the toilet
- Can be lower in cost than retrofitting plumbing from all the fixtures to capture greywater

CONS OF REUSING SEPTIC EFFLUENT

- Not typically suitable for installation by the average do-it-yourselfer
- Requires additional electricity use, and pumping of the water
- Not (yet) legal in some states
- Systems don't remove salts or boron from the water, which can harm plants; plant-friendly products must still be used in the house, and there is no way to turn "off" the system.

Using Treated Blackwater *for Irrigation and Fertilizing*

Jeremiah Kidd and his family reuse every last drop of water leaving the house in their landscape. Jeremiah is the owner of San Isidro Permaculture (see Resources), an ecological landscaping company specializing in greywater, rainwater, and edible landscapes in Santa Fe, New Mexico.

In his own home, Jeremiah decided to install a blackwater recycling system instead of the greywater systems he often installs. Why? The only growing areas on his land were uphill from the house, he'd have to pump the water with any reuse system, and the blackwater system gave him more options and control over the irrigation.

His home is plumbed conventionally; water from the shower, sinks, washing machine, and toilet flow together into the septic tank. The blackwater irrigation system begins in the second chamber of the tank with a SludgeHammer system (see Resources), pumping oxygen into the chamber to feed aerobic bacteria that clean the water. The treated septic effluent overflows into a pump tank where it's pumped out through a filter to subsurface drip irrigation designed especially for wastewater, using Netafim purple tubing for the irrigation lines. (New Mexico requires that tubing be buried 6 inches below grade.) Soil microbes further clean the water, and plants benefit from the nutrients.

Jeremiah's system has four zones: two zones of native plants, providing habitat for birds and beneficial insects; and two zones

Jeremiah's beautiful yard

"I think of this as a fertilizing system. It's not enough water for the entire landscape, but it reuses all the water from the home and sends nutrients to the plants."

of food production, food forest, and fruit trees. He concentrates the water in those zones during the growing season.

"I think of this as a fertilizing system. It's not enough water for the entire landscape, but it reuses all the water from the home and sends nutrients to the plants," says Jeremiah.

They are a water-conscious family, using less than 20 gallons per person per day. With a large landscape (they garden around 10,000 square feet) that amount of water covers only about one-third of the need. The other two-thirds is irrigated with rainwater and, occasionally, well water.

Jeremiah built the system himself (as well as the house). He spent around $4,000 for the parts. He was allowed to reduce his leach field significantly, from around 30 infiltrators down to 8. This is an accepted technology in New Mexico, and there were no problems with getting a permit. Maintenance is minimal: twice a year he cleans filters and once a year adds bacteria to the tank.

Treated blackwater irrigates and fertilizes the landscape.

Appendix

Plumbing Basics for Greywater Installation

To install your greywater system, you'll need to be familiar with basic plumbing techniques: namely cutting, assembling, and attaching pipes. You'll also need to know what tools to use and understand how the components fit together. The following section will go over common terminology for plumbing parts, how to cut and connect materials, and some basic installation tips. If you're new to plumbing, I recommend getting a general reference book (see Resources) for support and for help with tackling any obstacles you may encounter.

Parts Primer

Understanding a few basic terms will help you find the parts you need when you're building your system. Standard plumbing parts are identified by their size and material and how they connect together. Here we'll also cover some of the not-so-common valves that are regularly used in greywater installations, including check valves, backwater valves, and air admittance valves,

Types of Threads

Wouldn't it be nice if plumbing parts all had the same type of thread, and they all connected to one another? Unfortunately for the beginner, standard or "normal" threads don't exist. Let's look at the most common types of threads you'll encounter when installing a greywater system.

MALE VS. FEMALE. Male fittings have exposed threads; female fittings have internal threads. MPT = male pipe thread; FPT = female pipe thread.

PIPE THREAD VS. GARDEN HOSE THREAD (VS. BUTTRESS THREAD). NPT (national pipe thread) is the most common type of thread found on standard plumbing supplies. Pipe threads are different than the threads on a garden hose. You cannot connect a female-threaded garden hose to a fitting that has pipe threads. Buttress threads, or "coarse threads," are commonly used in 30- and 55-gallon drums and IBC totes (more commonly used in rainwater collection systems than in greywater systems). Buttress threads are not compatible with NPT or garden hose threads. (You can mail-order adapters for them, or connect to the NPT thread options on the containers).

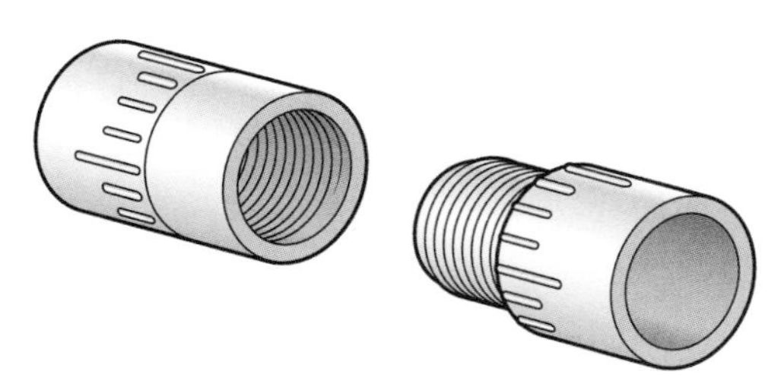

Female (left) and male threads

TAPERED VS. STRAIGHT THREADS. NPT threads are tapered. If you try to connect a male-threaded adapter into a female-threaded coupling, the tapers on the threads will prevent you from connecting them fully together. Plumbing fittings have tapered threads. For a watertight seal, tapered threads must be wrapped with pipe thread tape. Plastic electrical conduit fittings have straight threads, so it's possible to fully connect a male to female adapter. This is useful as a lower-cost way to create an outlet in a barrel (just add a rubber washer in between them for a watertight seal) and can be used instead of a tank adapter or bulkhead fitting.

Check Valves (Swing vs. Spring)

A check valve is a one-way valve used to prevent water from draining back down a pipe after it's been pumped out and uphill. For greywater applications, use **swing** check valves, not **spring**. Swing check valves have a flapper inside that easily pushes open, requiring less pressure than the spring check valves. Swing check valves are used in pumped systems, though not typically for the laundry-to-landscape, since most L2L system don't pump uphill. Don't use a check valve unless it's necessary, because it adds friction and is a potential clogging point. You can buy swing check valves at large irrigation or hardware stores.

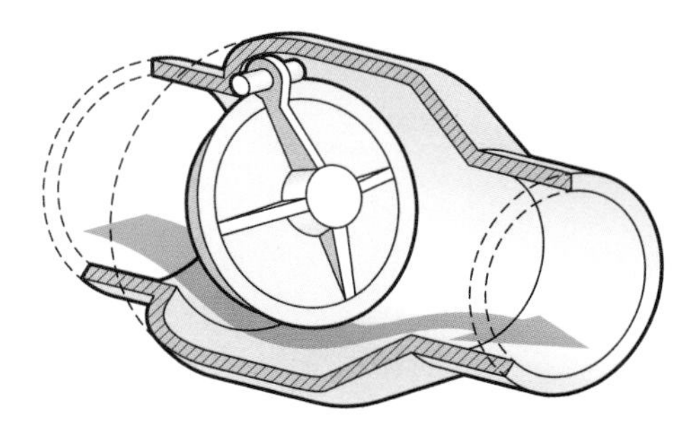

Swing check valve: water can flow in only one direction.

Air Admittance Valve (AAV)

An air admittance valve (AAV), also called an in-line vent or auto-vent, is a one-way mechanical vent used in plumbing to replace the need for a conventional vent pipe to the roof. AAVs allow air into the system to prevent water traps from siphoning. In an L2L system the AAV is used to prevent a potential siphon in the washing machine as it tries to refill. In this application, the AAV is not connected to the drainage plumbing of the house and is not being used to vent the washing machine drain. Note that plumbing codes typically require a specific type of AAV, made for venting a drain (which costs around $20 or $30), but in an L2L system any mechanical vent is suitable, including the lower-cost "auto-vent" option (around $4). See Resources for more info.

Note: An AAV should not leak unless it is defective or breaks. If water ever leaks from the vent, replace it.

Backwater Valve

A backwater valve is another type of one-way valve designed to prevent a sewer backflow from entering the surge tank or greywater line. Install one on the overflow pipe of a surge tank or on the sewer side of the pipe leading from a diverter valve. A backwater valve is used for gravity flow and is serviceable (you can unscrew it and open the valve for cleaning), unlike the swing check valve. Most backwater valves are installed horizontally, so plan accordingly. Plastic backwater valves are inexpensive (about $20 to $30) and are available from plumbing supply stores.

Union

A union fitting allows you to easily disconnect and reconnect the pipes. It can either be glued or threaded on a pipe. Install a union where you may want to disconnect the pipes; for example, to disconnect pipes to repair a pump. Unions are available at hardware and plumbing stores.

Transition Coupling

Transition couplings are used to connect different sizes and types of pipe together. The size and thickness of the rubber inside the transition coupling compensates for the size difference (outside diameter) between the pipes and forms a watertight seal between them. The easiest way to get the proper coupling is to go to a plumbing supply house and ask for help. You will need to know the pipe size (the inside diameter of the pipe) and the material of the two pipes you want to connect together. For example, to connect a 2-inch copper pipe to the diverter valve, which is 2-inch plastic, you'll need a 2-inch copper to 2-inch plastic transition coupling.

Use a transition coupling to connect your 3-way diverter valve into the greywater drain. First, loosen the bands holding the steel jacket over the rubber coupling. Then, slide the steel jacket over the existing pipe. Slide the rubber sleeve over the pipe, then the diverter valve. (Make sure to orient the rubber sleeve correctly; in this example the copper side must go onto the copper pipe, and the plastic side onto the plastic pipe.) You may need to roll back the rubber to fit the diverter in between the sections of rigid pipe, as shown. After the valve is in place, slide the metal sleeve over the rubber coupling and tighten the bands. Check this joint for leaks when you test your system (and tighten more if there is a leak).

INSTALLING A TRANSITION COUPLING

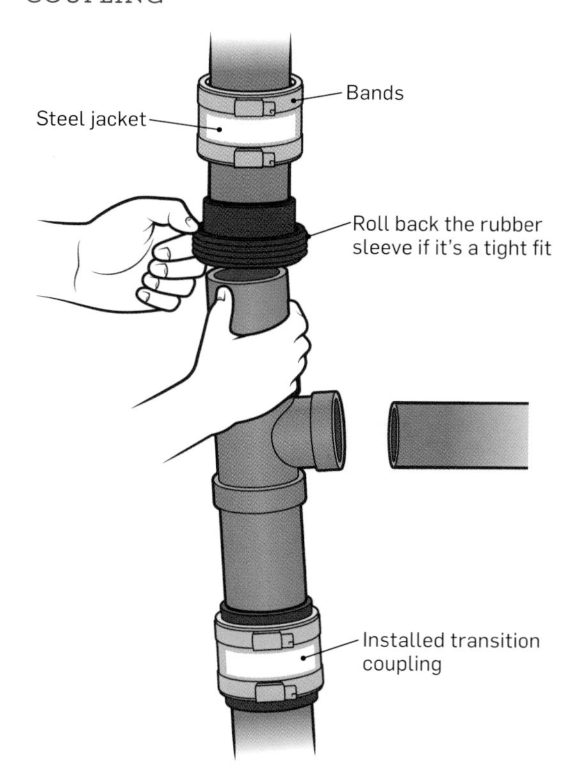

PIPE MATERIALS FOR DRAINAGE PLUMBING *(and how to work with them)*

Material	Options for Cutting	Connecting Joints
ABS (BLACK PLASTIC)	Tubing cutter with wheel for cutting plastic Handsaw for cutting plastic	Glue pipe to fittings with ABS cement
PVC (WHITE PLASTIC)	Tubing cutters with wheel for cutting plastic Large ratcheting cutters Handsaw for cutting plastic	Glue pipe to fittings with PVC primer and cement (or use Gorilla PVC, a less toxic self-priming cement)
GALVANIZED STEEL	Reciprocating saw with metal blade Grinder Hacksaw	Use appropriate transition coupling to connect to plastic pipe
CAST IRON	Grinder Chain-snap cutter Reciprocating saw with diamond blade	No-hub couplings (old cast-iron pipes were connected using "lead-and-oakum" joints)
COPPER	Tubing cutter with metal wheel Hacksaw	Use a transition coupling to connect copper pipe to plastic valve.

NOTES:

- Plastic pipe is the easiest and cheapest to work with. Interface between PVC and ABS with transition glue or a transition coupling.
- Steel pipe corrodes over time. Cut out as much old corroded pipe as possible and replumb. Be careful when cutting out a section of pipe; the vibrations can cause leaks in other pipe joints.
- With old cast iron plumbing, don't disturb lead-and-oakum joints. They're sealed with molten lead into the bell joint, packed with horsehair or jute. You may need to replace a section of plumbing.
- Copper is used for both water lines and drainage pipe. Water lines typically are 1 inch or smaller; drainage pipe is larger and has yellow markings. Don't accidentally cut into a water line!

Basic Installation Techniques

Following are some basic tips and guidance for connecting pipes to flow greywater out to the landscape. To learn more or to get hands-on training with plumbing basics, look for classes held at community colleges, home-improvement stores, or private schools; they can be a great place to learn tricks of the trade and gain tool confidence. Working with a knowledgeable friend is another way to learn your way around plumbing.

Connecting Threaded Fittings

To connect threaded fittings, wrap pipe thread tape around the threads; this helps create a watertight seal. Wrap clockwise, as you face the open end with the threads, overlapping at least three turns. If you wrap the wrong direction, the tape will come off when you screw in the fittings; rewrap if this occurs. Carefully screw the fitting in clockwise. A trick to make sure you are not "cross-threading" the fitting is to first screw in the fitting to the LEFT until you hear a "click" and feel the fitting settle into place. Then screw the fitting to the RIGHT to tighten it. It should thread easily at first, gradually getting harder and harder. Plastic threads are easy to damage when threading into metal, so go slow and easy.

Cutting and Gluing Plastic Pipe (PVC and ABS)

Plastic pipe is easy to cut and connect. To construct your greywater system, you'll need to know how to measure, cut, and glue the pieces together.

MEASURING. To measure a length of pipe, measure the distance between the fittings, as shown, then ADD the distance the pipe will slip into the fittings once it's glued (the glue lubricates the pipe so it slips into the fitting up to the lip). If you end up short, just cut the pipe and use a coupling to increase the length.

CUTTING. You can cut rigid plastic pipe with a simple handsaw (with a blade for plastic), though ratcheting cutters for 1-inch pipe and tubing cutters (with plastic wheel) for larger pipe make a nice, clean cut. Cuts from a handsaw create burrs, little balls of plastic around the cut; always remove these with fine sandpaper or scrape them off with a utility knife before gluing the pipe.

GLUING. Dry-fit the pipe and fittings together then mark the joints with ½-inch lines. When you glue, make sure the marks line up. Clean off the ends of the pipe with a rag before gluing. Connect PVC together using PVC solvent glue (and PVC primer, if required; check product label on the glue). Connect ABS pipes together using ABS cement. Use transition glue or transition couplings if you are connecting dissimilar plastics; this prevents leaks. Place old rags or newspaper on surfaces to protect them from dripping glue. Wear gloves.

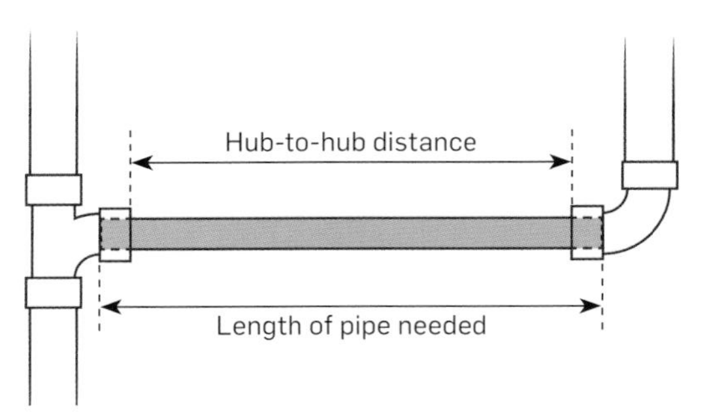

With glue, the pipe inserts fully into the fitting. Measure this length before cutting.

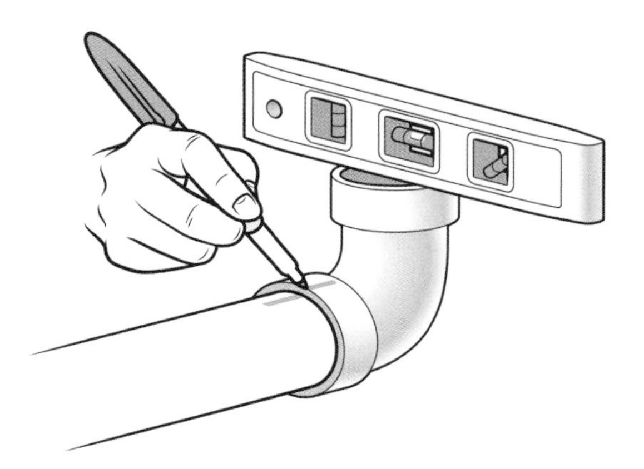

Mark pipe before gluing.

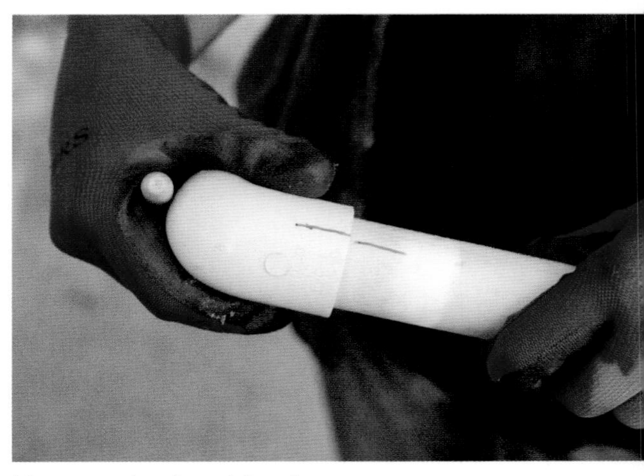

Line up marks when gluing pipe.

Apply glue on the inside of the fitting, then the outside of the pipe. Push the two together, lining up your marks. Hold for several seconds so the pipe doesn't push back out.

Tip: Gorilla PVC is less toxic than traditional PVC solvent glue and does not require primer like conventional solvent glue does.

Tips for Working with Irrigation Tubing

Uncoil the tubing the day before and leave it out in the sun. Try to work on a warm day, as the tubing will be softer and more pliable. To make it easier to fit tubing onto barbed fittings, dip the ends of the tubing in hot water to soften the plastic (bring a cup and thermos). In general, work with the natural curves in the tubing, and don't try to straighten it. To irrigate multiple plants along a straight line, alternate curved pieces to maintain an overall straight trajectory. Always lay the curves sideways, not up or down. Rotate tubing on the barbed fittings to change the tubing's orientation.

Using a Grade Level

A grade level is a level with a second set of lines around the bubble to show two-percent slope, standard drainage plumbing slope. It makes it easy to slope pipes: just align the bubble with the second line to find proper slope.

"Tool Tight"

In greywater systems fittings should be "tool tight." Too loose causes leaks, and too tight cracks plastic. First, tighten the fitting with your hands. Then, use large wrenches or tongue-and-groove pliers to tighten until it starts to become difficult to tighten more.

Tool Safety

Power tools make construction jobs faster but are dangerous if proper safety techniques aren't followed. Ask about tool safety if you're learning from someone else or using an unfamiliar tool. For example, if you're drilling a hole into wood and the drill bit gets stuck, the drill body will whip around and can smack you. With proper body positioning, a bruise is the worst that will happen. If you're unaware of where the drill may swing, you could be smacked in the face, knocked off a ladder, or get a broken wrist.

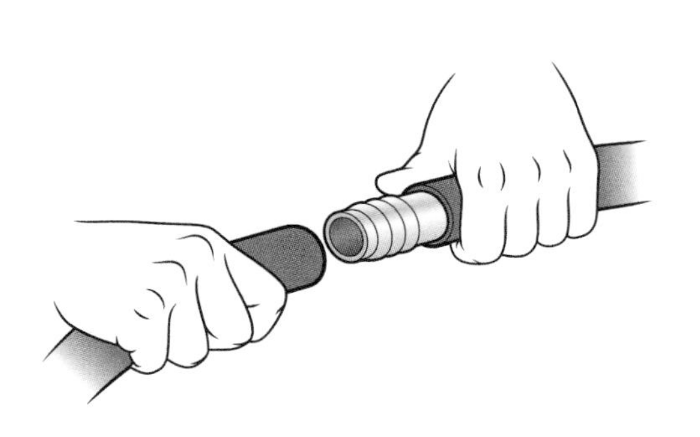

Hot water softens plastic. Dip the end of the tubing in hot water before pushing it over the barbed fitting ridges, so that it fits over the barbs with ease.

Resources

Planning Your Home Greywater System

A&L Western Laboratories, Inc.

www.al-labs-west.com

Soil laboratory offering low-cost soil-texture testing (for soil type)

Global Water Program

The Nature Conservancy

www.nature.org

Learn where your drinking water comes from

Greywater Action

www.greywateraction.org

List of greywater-compatible soaps and detergents

"A Guide to Estimating Irrigation Water Needs of Landscape Plantings in California"

www.water.ca.gov/wateruseefficiency/docs/wucols00.pdf

University of California Cooperative Extension and California Department of Water Resources, 2000. Document containing the species factor of many landscape plants and a reference for low-water-use plants. Access this information using the online tool at *www.waterwonks.us.*

Plant Finder

Sunset Magazine

http://plantfinder.sunset.com

Find water and soil pH requirements for specific plants in your region

Plumbing

DoItYourself.com

www.doityourself.com/scat/plumbing

Diagrams and instructions for how to stop leaks

San Francisco Water Power Sewer

http://sfwater.org/graywater

Greywater reuse information

SinkPositive

sinkpositive.com

Retrofit device for toilet lid that turns it into a sink

Skin Deep Cosmetics Database

Environmental Working Group

www.ewg.org/skindeep

Ingredients for personal care products

Surf Your Watershed

U.S. Environmental Protection Agency

http://cfpub.epa.gov/surf/locate

Find your watershed and connect with local watershed groups

Tracking Down the Roots of Our Sanitary Sewers

www.sewerhistory.org

California Urban Water Conservation Council

www.h2ouse.org

Resources for repairing all parts of the toilet, as well as many water-saving tips

Water Rebate and Incentive Programs

City of Tucson

www.tucsonaz.gov/water/rebate

WaterSense Program

U.S. Environmental Protection Agency

www.epa.gov/watersense

Find native and low-water-use plants for your region and connect with local native plant organizations, as well as an online tool for ET rates and average rainfall for peak irrigation month

Additional Reading

Allen, Laura, Bryan, S., and Woelfle-Erskine, C. "Residential Greywater Irrigation Systems in California: An Evaluation of Soil and Water Quality, User Satisfaction, and System Costs." Greywater Action, 2012.
Results of study of 83 residential greywater systems including water-saving data

Friedler, Eran and Roni Penn. "Study of the Effects of On-Site Greywater Reuse on Municipal Sewer Systems." The Grand Water Research Institute and the Technion Research and Development Foundation, 2011

Kuru, Bill and Mike Luettgen. "Investigation of Residential Water Reuse Technologies." Presentation at the Water-Smart Innovations Conference and Exposition, 2012
Study on 4 different toilet-flushing greywater systems

Ludwig, Art. *Create an Oasis with Greywater*, 5th ed. Oasis Design, 2009

———. *Water Storage: Tanks, Cisterns, Aquifers, and Ponds for Domestic Supply, Fire, and Emergency Use.* Oasis Design, 2005
Information on how to build a rainwater pond

Toensmeier, Eric. *Perennial Vegetables: From Artichoke to 'Zuiki' Taro, a Gardener's Guide to Over 100 Delicious, Easy-to-Grow Edibles*. Chelsea Green, 2007

Dell, Owen E. *Sustainable Landscaping for Dummies*. Wiley Publishing, 2009

Hemenway, Toby. *Gaia's Garden: A Guide to Home-Scale Permaculture*, 2nd ed. Chelsea Green Publishing, 2009

Lancaster, Brad. *Rainwater Harvesting for Drylands and Beyond*, Volumes 1 and 2 (2nd and 3rd reprints). Rainsource Press, 2013
Rainwater harvesting and passive solar design (using trees as natural air conditioning)

Allen, Laura. *The Water-Wise Home*. Storey Publishing, 2015
How to install rainwater systems

Ludwig, Art. *Create an Oasis with Greywater*, 5th ed. Oasis Design, 2009

Kourik, Robert. *Drip Irrigation: for Every Landscape and All Climates 2nd. ed.* Metamorphic Press, 2009.
Good overview of simple drip irrigation systems and how to install them

Building Your Greywater System

Aqua2Use Division
Matala Water Technology Co.
info@aqua2use.com
www.aqua2use.com
Manufactured greywater systems

Banjo Corporation
765-362-7367
www.banjocorp.com
Full port 3-way valve

Clean Water Components
www.cleanwatercomponents.com
Kits for building L2L and branched drain systems and actuators.

Dripworks
800-522-3747
www.dripworks.com
Fittings for L2L systems

Evergreen Lodge
Yosemite National Park
www.evergreenlodge.com
Simple and complex greywater systems

Fimco Manufacturing, Inc.
www.fimcomfg.com
Indexing valve for zoned irrigation; Wastewater Hydro Indexing Valve (10 psi)

GreyFlow Greywater Reuse Systems
www.greyflow.net.au
GreyFlow Plug-n-play system and GreyFlow Roto Valve

Greywater Action
www.greywateraction.org
Greywater reuse projects

Hydro-Rain
888-493-7672
www.hydrorain.com
Blu-Lock 1-inch irrigation tubing and fittings

Infiltrator Systems Inc.
800-221-4436
www.infiltratorsystems.com
Subsoil infiltrators

Legend Valve
800-752-2082
www.legendvalve.com
Full port 3-way valve

Morrow Water Systems
morrowwatersavers.com
Automated greywater systems

NSF International
www.nsf.org
Standard 50 certification for 3-way diverter valves

NutriCycle Systems
John Hanson
301-371-9172
http://nutricyclesystems.com
Installs subsoil infiltrators for greywater systems

Orenco Systems, Inc.
www.orenco.com
Indexing valve for zoned irrigation; mechanical distribution valve

Pentair Ltd.
www.pentairpool.com
Pentair 3-way diverter valve

PlumbingSupply.com
www.plumbingsupply.com
AAV and auto-vents

San Isidro Permaculture
Jeremiah Kidd
505-983-3841
http://sipermaculture.com

Sierra Watershed Progressive
www.sierrawatershedprogressive.com

SludgeHammer
800-426-3349
www.sludgehammer.net
Blackwater recycling system: septic system effluent to irrigation

WaterSprout
510-541-7278
www.watersprout.org
Designs and installs high-end greywater and rainwater systems

Zodiac International
www.zodiacpoolsystems.com
Jandy Space Saver diverter valve

Flow Splitters

HD Supply Maintenance Solutions
800-431-3000
http://hdsupplysolutions.com
Flow splitters (double ¼ bend)

Oasis Design
http://oasisdesign.net
Flow splitters with pre-installed threaded plug

Additional Reading

Creative Publishing International. *Black & Decker: The Complete Guide to Home Plumbing*, 5th ed. Cool Springs Press, 2012
——. *Black & Decker: The Complete Guide to Wiring*, 6th ed. Cool Springs Press, 2014

ACKNOWLEDGMENTS

This book developed out of the greywater chapters in my previous book, *The Water-Wise Home*. I want to extend my heartfelt thanks a second time to everyone who supported that book. There are so many of you! Thank you, all!

I want to give special thanks to Art Ludwig for his pioneering work on simple, low-tech greywater systems, and his original designs for the L2L and branched drain systems. His extensive research and resources have brought greywater information to so many people around the world.

Greywater Action. So much of the content and information in this book is composed of the shared knowledge of our group. Thank you to all the Greywater Action members past and present. It's been wonderful to work with my new friends and colleagues in LA. You all are awesome! Cris Sarabia, Ty Teissere, Laura Maher, and Sergio Scabuzzo. And the Bay Area crew: Christina Bertea, Brian Munson, Natalie Kilmer, and Jessica Arnett! Christina, thanks for the edits and helpful suggestions!

Big thanks to all the greywater installers who generously shared their experiences, photographs, and expertise with me, in particular Leigh Jerrard and Joe Madden (Greywater Corps), John Russel (Water Sprout), Regina Hirsch (Sierra Watershed Progressive), Sherry Bryan (Ecology Action), Alan Hackler (Bay Maples), and Allan Haskell (EnviroMeasure).

Thank you for the interviews and reviews: Bill Wilson, Jeremiah Kidd (San Isidro Permaculture), Steve Bilson (ReWater), Rob Kostlivy, Carl Warren, John Hanson (Nutricycle Systems), Paul James (Water ReNu), Paula Kehoe (SFPUC), Sam Milani (GreyFlow), and Bob Hitchner (Nexus Ewater).

Thank you to all the homeowners who let us come photograph your beautiful greywater systems. Eric and Peg, thanks for being such awesome hosts for the L2L install photo shoot! Leigh and Sergio, enormous thanks for all your help and support for the photo shoot days.

Thank you to the forward-thinking regulators out there — we need more of you! Thanks, Jeff Hutcher! Thanks, Rob Kostlivy! Thank you, Osama Younan! Thank you all at HCD!

Thank you Paddy Morrissey of Paddy Designs for your work on the original greywater-ready building images.

Thank you, Storey Publishing, for suggesting this book. Deb Burns, it is such a pleasure to work with you.

Big thanks to my parents for the endless support with all my projects.

And my family, Peter and Arlo, thank you for your love and support, with everything.

Index

Page numbers in *italic* indicate illustrations and photographs. Page numbers in **bold** indicate charts and tables.

Grow Your Green Library
with More Books from Storey

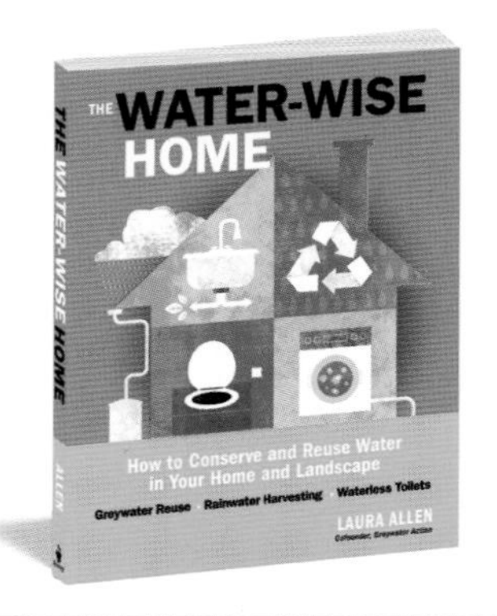

by Laura Allen

Learn how to conserve and reuse water throughout your house and landscape with this in-depth guide. Detailed illustrations and step-by-step instructions show you how to construct your own rain catchment system, composting toilet, and more.

by David A. Bainbridge

Tackle any drought with these inexpensive, low-tech methods for using water more efficiently in your garden. Illustrated, step-by-step instructions teach you to deliver water directly to roots, reduce weed growth, and harvest rainwater.

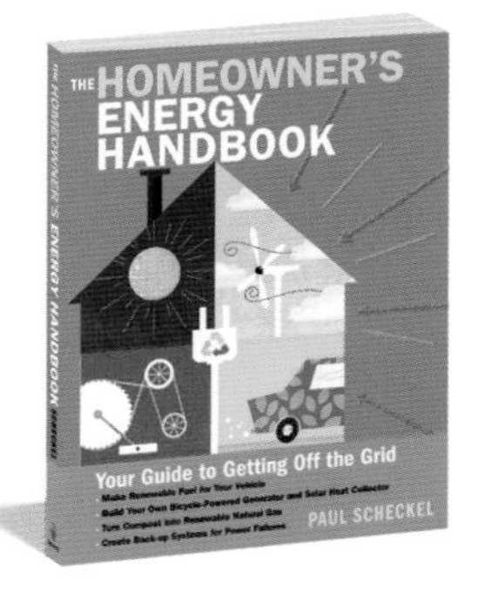

by Paul Scheckel

Whether you want to button up your house to be more energy-efficient, find deep energy savings, or take your home entirely off the grid, you'll get the knowledge and skills you need to reduce your use, then produce!

JOIN THE CONVERSATION. Share your experience with this book, learn more about Storey Publishing's authors, and read original essays and book excerpts at storey.com. Look for our books wherever quality books are sold or call 800-441-5700.